The Cape Town Convention

Studies in Space Law

VOLUME 8

The titles published in this series are listed at brill.com/slaw

The Cape Town Convention

Its Application to Space Assets and Relation to the Law of Outer Space

By

Mark J. Sundahl

MARTINUS NIJHOFF PUBLISHERS

LEIDEN · BOSTON
2013

Library of Congress Cataloging-in-Publication Data

Sundahl, Mark J.
The Cape Town Convention : its application to space assets and relation to the law of outer space / By Mark J. Sundahl.
pages cm. -- (Studies in space law)
Includes bibliographical references and index.
ISBN 978-90-04-20891-9 (hardback : alk. paper) 1. Convention on International Interests in Mobile Equipment (2001) 2. Aeronautics--Equipment and supplies 3. Space law. 4. Artificial satellites in telecommunication--Finance. 5. Contracts (International law) I. Title.

K4093.32001.S86 2013
341.4'7--dc23

2013025566

This publication has been typeset in the multilingual "Brill" typeface. With over 5,100 characters covering Latin, IPA, Greek, and Cyrillic, this typeface is especially suitable for use in the humanities. For more information, please see www.brill.com/brill-typeface.

ISSN 1871-7659
ISBN 978-90-04-20891-9 (hardback)
ISBN 978-90-04-25656-9 (e-book)

This book is printed on acid-free paper.

To my daughters,
Meredith Aracelia
and
Charlotte Hanna

Contents

Acknowledgements

The purpose of this page is to thank those whose influence and support have aided in the successful completion of this project. I must begin by thanking Drs. Tanja Masson-Zwaan who gave me entrée to the society of space lawyers when she invited me to speak on the Cape Town Convention at the IISL Regional Space Law Conference in Bangalore in 2005. There I had the good fortune to debate the merits of the Cape Town Convention with Paul Larsen, who is more qualified than I to write a book on the Cape Town Convention, and Dr. Rajeev Lochan who in 2007 left us too soon. I also met Prof. Frans von der Dunk for the first time in Bangalore and I am indebted to him for his steady encouragement and his invitation to publish a volume in this series. I am also indebted to Peter Nesgos who graciously allowed me to join the Space Assets Working Group for the Cape Town Convention where I had the opportunity to learn from the other participants in the drafting process, most memorably from Martin Stanford and Sir Roy Goode. This volume was enriched by the able research of Rachel Parks and Anna Brown whose assistance helped shed light on some of the contours of this study. The thoughtful comments of Prof. von der Dunk and an anonymous reviewer of the draft manuscript raised my work to a higher level. The patience of Ms. Bea Timmer at Brill allowed me the time to refine my writing. I am also grateful to Dr. Rosa DelVecchio, who proofread my drafts and made many improvements. Finally, I would like to thank Unidroit and the International Civil Aviation Organization for giving their permission to include in this volume the text of the Cape Town Convention and the Space Assets Protocol. Any errors or shortcomings herein remain entirely my own.

Those of us who work in the field of space law think frequently of what is to come. With these dreams of the future, I dedicate this book to my daughters, Meredith Aracelia and Charlotte Hanna.

Mark J. Sundahl

List of Abbreviations

Cape Town Convention	Convention on International Interests in Mobile Equipment
C.F.R.	United States Code of Federal Regulations
Declaration of Principles	Declaration of Legal Principles Governing the Activities of States in the Exploration and Use of Outer Space
Draft Articles	Draft Articles on Responsibility of States for Internationally Wrongful Acts
Liability Convention	Convention on International Liability for Damage Caused by Space Objects
Moon Agreement	Agreement Governing Activities of States on the Moon and Other Celestial Bodies
Outer Space Treaty	Treaty on Principles Governing the Activities of States in the Exploration and Use of Outer Space, Including the Moon and Other Celestial Bodies
Registration Convention	Convention on Registration of Objects Launched into Outer Space
Rescue and Return Agreement	Agreement on the Rescue of Astronauts, the Return of Astronauts and the Return of Objects Launched into Space
Space Assets Protocol	Protocol to the Convention on International Interests in Mobile Equipment on Matters Specific to Space Assets
U.C.C.	Uniform Commercial Code (The American Law Institute and the National Conference of Commissioners on Uniform State Laws)
U.S.C.	United States Code

Chapter One

Introduction

The adoption of the Cape Town Convention on International Interests in Mobile Equipment (the "Cape Town Convention") in 2001 marked a significant step forward in the international law of asset-backed finance.[1] The Cape Town Convention created a new international regime for financing assets in the aviation, rail transportation, and space industries in order to facilitate investment in these capital-intensive sectors. The Convention's entry into force with respect to aircraft in 2006 transformed the financing of airline fleets.[2] Now the extension of the Cape Town Convention to transactions involving space assets by way of the Protocol to the Convention on International Interests in Mobile Equipment on Matters Specific to Space Assets (the "Space Assets Protocol") promises to be a transformative event in the field of space law as well.[3] In addition to bringing international harmonization to the law governing the financing of satellites and other space assets, the Convention's application to space assets raises complex issues regarding the relationship between the Convention and the existing law of outer space. Space law is a multi-faceted field of law that covers a variety of issues addressed in five United Nations treaties. In addition to these treaties, the Constitution and Convention of the International Telecommunications Union comprise a further body of international law that governs the use of orbital positions and radio frequencies – a

---

1 Convention on International Interests in Mobile Equipment, Nov. 16, 2001, S. Treaty Doc. No. 108-10, available at www.unidroit.org/english/conventions/c-main.htm [hereinafter Cape Town Convention].

2 See *infra* Section 1.5 regarding the application of the Cape Town Convention to aircraft.

3 Protocol to the Convention on International Interests in Mobile Equipment on Matters Specific to Space Assets, in *Final Act of the Diplomatic Conference for the Adoption of the Draft Protocol to the Convention on International Interests in Mobile Equipment on Matters Specific to Space Assets*, UNIDROIT Doc., DCME-SP–Doc. 43 (9 March 2012), available at www.unidroit.org/english/workprogramme/study072/spaceprotocol/conference/documents/dcme-sp-43-e.pdf [hereinafter Space Assets Protocol].

regulatory system that is central to the orderly operation of the global telecommunications industry. Multiple aspects of space law and telecommunications law are implicated by the Cape Town Convention and the Space Assets Protocol. In some cases, these intersections of law create potential conflicts that may be problematic to the practitioner. In other cases, the interaction between these different systems of law creates new synergies that may result in the more effective operation of the law.

This book is designed not only to explain the technical operation of the Convention and its application to space assets by way of the Space Assets Protocol, but also to explore the many intersections between the Convention and the existing body of space law. The practitioner involved in space finance, whether a lawyer, banker, or business executive, will find herein a clear explication of how the Cape Town Convention will affect their transactions.[4] The legal academic with a broader interest in the place

[4] Practitioners should also consult the official commentaries to the Cape Town Convention and protocols written by Prof. Sir Roy Goode. Goode, R., *Convention on International Interests in Mobile Equipment and Protocol Thereto on Matters Specific to Aircraft Equipment: Official Commentary* (2nd ed. 2008) [hereinafter Official Commentary]. The official commentary on the Space Assets Protocol had not yet been published at the time this book was written. The Unidroit website (www.unidroit.org) also contains a rich trove of resources including the texts of the Convention and protocols, the preparatory work for these instruments, and bibliographies that collect the writings on the Convention and protocols. Some of the fundamental works on the nature, history, and operation of the Cape Town Convention include Stanford, M.J., "The Availability of a New Form of Financing for Commercial Space Activities: The Extension of the Cape Town Convention to Space Assets", 1 Cape Town Convention J. 109 (2012) [hereinafter Stanford 2012]; Stanford, M.J., "Completion of a First Draft of UNIDROIT's Planned Future Convention on International Interests in Mobile Equipment", 1 Unif. L. Rev./Revue de droit uniforme 274 (1996); Stanford, M.J., "A Broader or Narrower Band of Beneficiaries for the Proposed New International Regimen?: Some Reflections on the Merits of the Convention/Protocol Structure in Facilitating the Former", 4 Unif. L. Rev./Revue de droit uniforme 242 (1999) [hereinafter Stanford 1999]; Goode, R., "Transcending the Boundaries of Earth and Space: The Preliminary Draft UNIDROIT Convention on International Interests in Mobile Equipment", 3 Unif. L. Rev./Revue de droit uniforme 52 (1998); Clark, L.S., "The 2001 Cape Town Convention on International Interests in Mobile Equipment and Aircraft Equipment Protocol: Internationalising Asset-Based Financing Principles for the Acquisition of Aircraft and Engines", 69 J. Air L. & Com. 3 (2004); Cuming, R.C.C., "The Draft UNIDROIT Convention on International Interests in Mobile Equipment", 30 UCC L.J. 365 (1998); Cuming, R.C.C., "'Hot Issues' in the Development of the (Draft) Convention International Interests in Mobile Equipment and the (Draft) Aircraft Equipment Protocol", 34 Int'l Law. 1093 (2000) [Cuming 2000]; Wool, J., "The Next Generation of

of the Cape Town Convention in the body of space law will find a thorough exploration of the interplay of the Convention and existing space treaties. However, the divide between the practical and the academic – much like the divide between air space and outer space – is not sharp. The practitioner cannot analyze and apply the Convention and Protocol in a vacuum. The intersection of these instruments with existing space law requires that the practitioner understand how space law may affect the transaction at hand. In some cases, space law may complicate issues under the Convention. However, certain provisions of the space treaties may also in some cases provide helpful tools for the practitioner. This book identifies the different ways in which space law can affect transactions under the Cape Town Convention, while also examining the impact of the Cape Town Convention on the existing body of space law.

1.1. *An Overview of the Financing of Space Ventures*

The construction, launch, and operation of space assets require tremendous financial resources.[5] The construction and launch of a single satellite

---

International Aviation Finance Law: An Overview of the Proposed UNIDROIT Convention on International Interests in Mobile Equipment as Applied to Aircraft Equipment", 20 U. Pa. J. Int'l Econ. L. 499 (1999); Wool, J., "Rethinking the Notion of Uniformity in the Drafting of International Commercial Law: A Preliminary Proposal for the Development of a Policy-based Unification Model", 2 Unif. L. Rev./Revue de droit uniforme 46 (1997); Chinkin, C., and Kessedjian, C., "The Legal Relationship Between the Proposed UNIDROIT Convention and Its Equipment-specific Protocols", 4 Unif. L. Rev/Revue de droit uniforme 323 (1999). Additional works are listed in the bibliography.

5 Although the commercial space sector is typically understood as including primarily those satellite operators that provide space-based services such as satellite communications services, remote sensing and earth observation, and satellite navigation services, the industry embraces a variety of other companies and service providers that make the provision of these services possible, such as launch service providers, satellite manufacturers, financiers, and insurance brokers and underwriters. Even in the wake of the global financial crisis of 2008, the commercial space sector has recently seen continuous growth in revenues. Stanford 2012, *supra* note 4, at 117. In 2011, the commercial space sector's global revenues grew by 12% to a total of US$289.77 billion. Id. Much of this growth has been attributed to the growth within the commercial space infrastructure sector as the demand for personal navigation devices is growing in developing and emerging markets. Id. Continued expansion of the commercial space sector is expected for the near future. A considerable amount of growth is expected in emerging markets with a growing

can easily run into the hundreds of millions of dollars and the value of a single transponder on a satellite can range from five to twenty million dollars.[6] The types of transactions that enable the financing of these hyper-expensive space ventures vary. Equity finance, secured and unsecured lending, and project finance structures have all been utilized to raise capital. Equity finance refers to the raising of funds by the sale of a company's shares of stock. This stock can be sold through public offerings in a stock exchange or through private offerings to individuals or companies. Funds can also be raised by borrowing money from a bank or a syndicate of banks. Such loans can be unsecured so that banks have no right to the borrower's assets if the borrower fails to repay or, more typically, the loans can be secured on the assets of the borrower in order to provide the banks with some protection in the event of default.

Project finance is a more complicated method of financing capital-intensive ventures in which the lenders financing the project rely on the revenue generated by the project for repayment of the debt obligations without recourse to the company sponsoring the project. This structure appeals to companies because it limits the company's potential losses to the value of the assets involved in the project. In the case of satellite finance, lenders would look to the income generated from transponder leases and other revenue streams flowing to the satellite operator. Because the lender does not have recourse to other assets of the operator if the operator defaults on the loan obligation, the lender will demand a first-priority security interest in the satellite since this collateral is the lender's only protection in the event that the expected revenue streams do not materialize.[7]

Although satellite operators typically have their satellites built to their specifications, satellites have also been sold or leased to a new operator while on-orbit.[8] Individual transponders on a satellite are also commonly

---

demand in transponders and bandwidth. United States Federal Aviation Administration Commercial Space Transportation and the Commercial Space Transportation Advisory Committee, *2012 Commercial Space Transportation Forecasts* 17–18 (May 2012).

6 Nesgos, P., "Satellites and Transponders" §30.02.1, in Wong, J. and Dubin, B. (eds.) *Equipment Leasing* (2012).

7 For a thorough treatment of the complexities of project finance see Yescombe, E.R., *Principles of Project Finance* (Academic Press 2002).

8 A representative example of a satellite leasing took place in 2002 when the company SES S.A., a Luxembourg company, leased the Astra-1B satellite to the Swedish company Nordic Satellite AB. See Gerhard, M., "Transfer of Operation and Control with Respect to Space Objects – Problems with Responsibility and Liability of States", 51 *Zeitschrift für Luft- und*

sold and leased.[9] In some cases, a company will simply purchase or lease a transponder from the satellite operator, while in other cases a company may utilize a sale/leaseback structure to acquire the long-term use of a transponder.[10]

The legal environment in which these transactions take place is a major factor in ensuring that the transactions can be structured in a manner that reduces costs and risks to a level that is acceptable to the parties involved. In particular, the laws applicable to the enforcement of security interests are critical in a financing involving secured debt and the laws governing leasing and sales will affect the success of these types of transactions.

At the outset of the space age, governments funded their own space activity.[11] When private industry entered the satellite telecommunications field, the companies were able to finance their operations without relying on debt finance secured on their space assets. These were blue-chip companies with triple-A ratings and significant assets on the ground, such as AMERICOM (a subsidiary of RCA), SES, and DirectTV (a joint venture of RCA and Hughes Electronics). These companies had little difficulty receiving loans on the basis of their own creditworthiness and their easily accessible and marketable terrestrial collateral.[12]

Due in part to the deregulation of satellite communications and the opening of domestic markets to foreign satellite operators through the World Trade Organization (WTO) in the 1990s, new entrants into the

Weltraumrecht 571, 573 (2002). For additional examples of satellite leases see von der Dunk, F.G., "The Illogical Link: Launching, Liability and Leasing", in *Proceedings of the Forty-Sixth Colloquium on the Law of Outer Space* 349 (1993).

9 Transponders are more commonly subject to leases than satellites as a whole due, in part, to the fact that the operation of a transponder alone is not subject to licensing requirements. See Nesgos, *supra* note 6, §§30.02.1; 30.06.1.

10 Sale/leaseback financings involve the initial purchase of a transponder from the satellite operator by the company, which then resells the transponder to a bank. The bank then leases the transponder to the company. For an example of a Transponder Purchase Agreement see Hermida, J., *Transponder Purchase Agreement*, in *International Business Transactions: Standard Forms and Documents*, (Campbell D., ed., Kluwer Law International 2007; Suppl. 55 – May 2012).

11 For an overview of the government funding of space activity see Diedericks-Verschoor, I.H. Ph., "The Development of Financing of Spacecraft", in *Proceedings of the Fortieth Colloquium on the Law of Outer Space* 212 (1998).

12 Larsen, P.B., and Heilbock, J.A., "UNIDROIT Project on Security Interests: How the Project Affects Space Objects", 64 J. Air L. & Com. 703, 706 (1999).

satellite telecommunications industries began to emerge.[13] In 1996, the conclusion of the Fourth Protocol to the General Agreement on Trade in Services opened the domestic telecommunications industry of participating member states to foreign market entrants on a "most favored nation" basis.[14] This means that a state that is a member of the WTO cannot treat one state more favorably than another with respect to granting entry into its domestic telecommunications market, but must accord the same right to enter its market to every state (provided that the state has itself made a commitment to the Fourth Protocol).[15] This had the effect of opening domestic markets of the WTO members to telecommunication satellite operators in other countries – and these newly opened markets created opportunities for the expansion of the telecommunication satellite industry. More recently, a new cadre of space entrepreneurs have started companies in the field of both manned and unmanned space flight.[16] Unlike the established behemoths of the space industry, these new companies do not have a long history of creditworthiness or the vast assemblage of assets on the ground. These startup companies bring a new level of risk for financiers who intend to extend loans to finance their operations – and the assets available as collateral are limited to the company's satellites in orbit. These smaller companies will therefore have no choice but to use their space assets as collateral and the ability of financial institutions to acquire an enforceable security interest in these space assets will be essential to attract financing.[17] In addition, developing countries seeking the benefits of space

13 Lyall, F., and Larsen, P.B., *Space Law: A Treatise* 443 (2009); Davis, S.A., "Unifying the Final Frontier: Space Industry Financing Reform", 106 Commercial Law Journal 455, 457 (2001).

14 Fourth Protocol to the General Agreement on Trade in Services, S/L/20 (Apr. 30, 1996), 36 I.L.M. 366 (1997). See, *generally*, Meisner, S.M., "Global Telecommunications Competition a Reality: United States Complies with WTO Pact", 13 American Univ. Int'l L. Rev. 1345 (1998).

15 Meisner, *supra* note 14, at 1353.

16 Companies such as SpaceX, SpaceHab, Rocketplane Kistler, PlanetSpace, and SpaceDev have competed for an opportunity to provide cargo and crew delivery services for NASA under the Commercial Orbital Transportation Services (COTS) program. Regarding the COTS program see *NASA Commercial Crew & Cargo Program Office Homepage*, available at www.nasa.gov/offices/c3po/home/index.html. Suborbital space tourism companies, including Virgin Galactic and XCOR Aerospace, have also entered the field of space flight.

17 Panahy, D.A., and Mittal, R., "The Prospective UNIDROIT Convention on International Interests in Mobile Equipment as Applied to Space Property", 4 Unif. L. Rev./Revue de

may not have access to capital required to build and launch satellites and these countries may therefore benefit from the ability to leverage their space assets as collateral to attract private funding.[18]

1.2. *The Challenges of Financing Space Assets*

While secured loans have been used to finance space ventures, these loan obligations have generally been secured not on space assets, but on more

---

droit uniforme 303, 305 (1999); Stanford, M.J., and Poulain, B., "The Preliminary Draft Space Assets Protocol to the Convention on International Interests in Mobile Equipment on Matters Specific to Space Assets", in *Proceedings of the United Nations/Thailand/ European Space Agency Workshop on Space Law: Activities of States in Outer Space in Light of New Developments: Meeting International Responsibilities and Establishing National Legal and Policy Frameworks, Bangkok, 16–19 November 2010* at 5 (2010); Burman, H., "UNIDROIT Mobile Equipment Convention: Draft Protocol on Space Assets", in Cummins, S.J. (ed.), *2005 Dig. U.S. Practice in International Law* 747 (2005); Hazan, N., "The UNIDROIT Preliminary Draft Protocol on Matters Specific to Space Assets", 28 Annals Air & Space L. 219, 222 (2003); Stanford, M.J., "The UNIDROIT Project for the Creation of a New Legal Regimen Governing the Taking of Security in High-Value Mobile Assets: A Window of Opportunity for the Commercialisation of Space", in *Proceedings of the Third ECSL Colloquium: International Organisations and Space Law, Perugia, Italy, 6–7 May, 1999* 375, 378 (1999); Stanford, M.J., and Fontmichel, A. de, "Overview of the Current Situation Regarding the Preliminary Draft Space Property Protocol and Its Examination by COPUOS", 6 Unif. L. Rev./ Revue de droit uniforme 60, 62 (2001); Davis, *supra* note 13, at 458; Larsen and Heilbock, *supra* note 12, at 707 (explaining that "[n]ew privately-operated satellite networks require vendor finance form the manufacturers or project finance by institutional lenders" and that improved protection of lender interests "would facilitate space commerce."); UNIDROIT 1996, Study LXXII – Doc. 26, Add.2.

18 dos Santos, A.F., "Financing of Space Assets", 19 Space Policy 127, 127 (2003); Lochan, R., "Cape Town Convention & Space Protocol: A Critical Analysis", in Gopalakrishnan, V. and Lochan, R. (eds.), *Conference Proceedings of the ISRO/IISL Space Law Conference 2005: Bringing Space Benefits to the Asian Region*, 6–39, 6–40 (2006) (arguing that the Convention is needed by developing countries that "find funding of space projects beyond their reach"); Ribbeslink, O.M., "The Protocol on Matters Specific to Space Assets", 12 European Rev. Private L. 37, 38 (2004). Governmental need for external financing also exists to some extent in the industrialized world. See, e.g., Larsen and Heilbock, *supra* note 12, at 737 (explaining that Russia supported a broad interpretation of the assets to be covered by the Cape Town Convention because the Russian space industry "needs to be financed by foreign sources."); Comments by the Cosmic Space Agency of the Russian Federation, UNIDROIT 1996, Study LXXII – Doc. 26, at 1.

easily accessible, immobile, and marketable terrestrial collateral.[19] In order to comprehend why this is the case, the basic theory of an efficient system of secured credit must be understood. A predictable and efficient law of secured transactions reduces the cost of borrowing money because banks will charge a lower interest rate when they face a lower risk of financial loss. Without a predictable law of secured transactions, a creditor will not have this confidence and will be reluctant to release funds or, if the loan is issued, the bank will charge a higher interest rate.

At the most basic level, an efficient law of secured transactions will contain certain fundamental features. First, a lender must be able to easily acquire a non-possessory security interest without burdensome formalities. The need for a non-possessory security interest is self-evident. If the lender has to take possession of the collateral in order to have a valid security interest (or in order to perfect its security interest to ensure the priority of its claim over other competing claimants), the debtor will not be able to use the collateral. For some jurisdictions, the idea of a non-possessory security interest is (surprisingly) still a somewhat novel concept. It is also important to allow for the creation of a security interest without excessive formalities in order to keep transactional costs low, but also to avoid the possibility of a lender's security interest later being challenged for lack of compliance with certain formalities. For example, the requirements for describing the collateral in a security agreement should be lax. The greater specificity required in the description of the collateral, the greater the likelihood that the lender will not provide the requisite information (or perhaps make an error in, for example, one of the digits of a serial number).

Second, an efficient law of secured transactions must have clear priority rules coupled with a publicly searchable registration system to clarify the priority positions of competing claimants. A prospective lender must be able to ascertain that it will have priority over other subsequent lenders that may also take a security interest in the same collateral (as well as over other competing claimants such as a party that wins a court judgment against the debtor and collects its award by seizing the debtor's collateral). If the prospective lender does not have this certainty, it will not disburse the loan. Such certainty can be achieved through the adoption of priority

[19] One example of a satellite finance structure utilizing a security interest in the satellite involved the APSTAR satellite. See Panahy and Mittal, *supra* note 17, at 305. ProtoStar Ltd. also granted security interests in its satellites to an affiliate of Credit Suisse Group

rules that give priority to the party that registers its security first in a centralized publicly searchable registry. With this system in place, a prospective lender can simply search the registry to see if any security interests have been registered. If not, the lender will be the first to register and can be assured of first priority. In connection with this, the law should also allow for prospective registration (that is, the registration of the security interest before the loan is disbursed).

Third, a creditor must be able to unilaterally exercise remedies in the event of the debtor's default without the need to seek a court order. Litigation is expensive and any costs expended by the creditor in the process taking control of the collateral after the debtor defaults and selling the collateral (or exercising other remedies) deprives the creditor of funds that it could have otherwise applied directly to discharging the debtor's obligation. The creditor might be able to recoup the funds spent on such litigation upon the sale of the collateral, but if the collateral has limited value then the proceeds of the sale may not be sufficient to cover both the obligation owed to the creditor as well as litigation costs. In addition, the delay that results from litigation can result in the diminution of the value of the collateral. Technology, including satellite technology, becomes rapidly obsolete and, if litigation drags on for years, the collateral may have significantly less value when the creditor finally has the right to sell it.

Finally, a creditor must be able to exercise remedies against the debtor's assets even in the context of the debtor's insolvency. The primary purpose of bankruptcy proceedings is to protect the assets of the bankrupt party (by placing a stay on the actions of creditors that want to seize the bankrupt party's assets) and then to reorganize the company, discharge the bankrupt party's debts, and distribute assets to creditors in accordance with varying methods. If a secured creditor risks losing the collateral should the debtor enter bankruptcy, the security interest has little value and provides the creditor with no protection against the risk of default. Therefore, it is critical that a secured creditor be able to overcome any stay placed on actions against the debtor's assets and enforce its security interest against the collateral.

In addition to these core features, an efficient law of secured transactions also benefits from additional provisions that either ease the transactional burdens on the creditor or increase the creditor's ability to reach

AG in connection with the financing of its venture. See Michael Bathon, *ProtoStar Wins Court Approval of Liquidation Plan*, available at www.bloomberg.com/news/2010-10-06/protostar-wins-court-approval-of-sale-plan-resolving-creditors-disputes.html.

collateral in the event of default. For example, creditors benefit from a law that permits the creation of security interests that secure "future advances" in both existing and "after-acquired" collateral. In other words, a law with these provisions allows a security interest to secure not only the initial loan given to the debtor, but all subsequent loans (or "future advances"). This eliminates the need to enter into subsequent agreements when subsequent loans are made. Similarly, the transactional burden is eased if the security interest attaches not only to collateral in existence at the time the security agreement is entered into, but also to collateral that is acquired in the future (in the event that, for example, additional satellites are placed into orbit in a telecommunications satellite constellation). Greater protection against the risk of the debtor's default is also given to the creditor if the creditor's security interest continues in the collateral even if the debtor sells it to a third party and if the security interest automatically attaches to any proceeds realized by the debtor upon the sale of the collateral (or upon the destruction of the collateral). Thus, if the debtor sells a satellite to another company, the creditor would still be able to enforce its security interest against the satellite in the hands of the seller and would have a right to any funds received by the debtor upon the sale of the satellite.

In addition to having the protections of the proper legal environment, banks also need to have confidence that the collateral securing its loans will be easily marketable so that the bank can quickly dispose of the collateral and realize the proceeds needed to cover its losses.[20] A security interest in immobile terrestrial assets that are held in a country with an efficient and creditor-friendly law of secured transactions, such as the United States, provides lenders with a highly reliable security package. In contrast, lenders may lack confidence in security interests in space assets for a number of reasons.

First, the patchwork of domestic laws that may be implicated in the course of creating and enforcing a security interest in space assets can result in legal uncertainty and increase transactional costs.[21] Although

20 For a more detailed explanation of the benefits of security interests see McCormack, G., *Secured Credit Under English and American Law* 1–38 (2004).

21 See Buxbaum, H.L., "Unification of the Law Governing Secured Transactions: Progress and Prospects for Reform", 8 *Unif. L. Rev./Revue de droit uniforme* 321, 323–324 (2003); Larsen, P.B., "Creditors' Secured Interests in Satellites", in *Proceedings of the Thirty-Fourth Colloquium on the Law of Outer Space* 233, 236 (1991). Regarding the complexities of cross-border secured transactions see also Cuming, R.C.C, "Study of International Regulation of Aspects of Security Interests in Mobile Equipment", 1 *Unif. L. Rev/Revue de*

some countries, such as the United States, Australia, and New Zealand, have adopted progressive, creditor-friendly secured transactions statutes, many jurisdictions around the world retain an approach to secured transactions that fails to incorporate the above-mentioned fundamental principles of asset-backed finance.[22] For example, German law does not generally provide for the public registration of security interests for purposes of establishing priority and giving notice to prospective creditors of existing security interests.[23] Instead, priority in Germany is given to competing secured parties in order of the time of the creation of their security interests.[24] This can result in hidden liens if the debtor does not disclose existing security interests to a prospective lender since there is no public record of existing security interests. The laws of certain states do not allow secured parties to prospectively register their security interests, that is, to register their security interests before creating the security interest and issuing the loan. This places the secured party in jeopardy because the priority may not be assured (through registration) until after the loan is disbursed.[25] Some jurisdictions do not allow for a security interest in the original collateral to attach to the proceeds realized upon the debtor's disposition (or the destruction) of the collateral.[26] This prevents a secured party from asserting a right to receive, for example, the insurance proceeds paid out to the debtor in the event that the original collateral was accidently destroyed, thus placing the risk of loss entirely on the secured party. In India, although judge-made law gives creditors the nominal right to enforce

---

droit uniforme 63 (1990); Akseli, O., *International Secured Transactions Law: Facilitation of Credit and International Conventions and Instruments* (2011); McCormack, G., *Secured Credit and the Harmonization of Law: The UNCITRAL Experience* (2011); and Dahan, F. and Simpson, J. (eds.), *Secured Transactions Reform and Access to Credit* (2008).

22 Regarding the detrimental aspects of the law of secured transactions in Napoleonic jurisdictions (i.e., civil law jurisdictions that have enacted laws based on the traditions of the French Civil Code) see Wood, P.R., *Comparative Law of Security Interests and Title Finance* §23-016 (2007) (stating that the rules in these jurisdictions tend to be debtor-protective with an emphasis on judicial oversight of the enforcement of security interests in addition to other limitations).

23 See Hesdahl, R. and Lange, A., *Germany* §15:23, in *International Secured Transactions* (Campbell, D., ed. Thomson Reuters 2012).

24 See id. §15:25.

25 Fleisig, H., Safavian, M. & de la Pena, N., *Reforming Collateral Laws to Expand Access to Finance* 40 (2006).

26 Id. at 33–34.

a non-possessory security interest without a court order, the reality is that a court order is still generally required and the ensuing litigation can cause sufficient delay so as to destroy the practical value of a security interest.[27] This difficulty in creating and enforcing security interests in personal property is a problematic area of commercial law not only in India, but in many other developing countries as well.[28]

Second, although there is a general movement towards allowing parties to choose the law that governs their transaction (and thereby avoid inimical laws), parties that are competing for the priority of their security interests in a space asset will not have entered into any agreement in which they would have chosen applicable law. In this case, the conflicts of law rules of the forum where the competing claimants appear in court will determine the applicable law. The conflicts of law rule that generally applies to issues regarding the priority of security interests will, in many jurisdictions, call for the application of the law of the place where the collateral is located (*lex rei sitae*).[29] Traditionally, the application of this rule provided predictability for parties that sought to assert rights over the property at issue. However, when dealing with highly mobile collateral that frequently crosses borders, the predictability is lost because the location of the collateral can change without notice. The *lex rei sitae* rule is particularly problematic in the case of satellites and other space assets since assets in space are not located within the territory of any state.[30] The law applied by the court in this situation may result in a secured party unexpectedly losing

---

[27] Asian Development Bank, Regional Technical Assistance Project No. 5795-Reg, Insolvency Law Reforms, Report on India, Section C, available at http://www.insolvencyasia.com/insolvency_law_ regimes/india/section_c.html.

[28] See, generally, de la *Peña*, N., "Reforming the Legal Framework for Security Interests in Mobile Property", 4 Unif. L. Rev./Revue de droit uniforme 347 (1999). See also Panahy and Mittal, *supra* note 17, at 305 (citing, as an example, the uncertainty of Indonesian law regarding the use by Asia Cellular Satellite of its GARUDA satellite as security for debt which was effected by the transfer of the satellite to a security agent by means of a fiduciary transfer agreement).

[29] In the United States, the Uniform Commercial Code provides for parties to choose the applicable law in general, but requires that the law of the location of the collateral governs issues of priority. See *infra* Section 1.3. Another example of the application of the *lex rei sitae* is found in Italian law which requires that *in rem* rights in movable assets be governed by the law of the jurisdiction in which the assets are located. See Nunziante U., *Italy* §20:42, in *International Secured Transactions* (Campbell, D., ed. Thomson Reuters 2012) (referencing Article 51 of Law Number 218/1995).

[30] See *infra* Section 1.3.

priority. Finally, even if the parties to a dispute have chosen the governing law or the choice of law rules of the forum call for the application of the law of another jurisdiction, certain mandatory rules of the forum (regarding for example the procedure for enforcing security interests) may still be applied by the court.[31] In addition to creating legal uncertainty that could harm a secured party's interests, this uncertainty increases transactional costs due to the expense of engaging local counsel in multiple jurisdictions to ensure that the requirements for a valid security interest, and the perfection of that security interest, are met.[32]

Third, the lack of a centralized international registry of security interests to establish clear priority of competing claimants creates uncertainty for creditors regarding the priority of their security interests.[33] Some countries even have separate registries for each of their territorial units (as is true in the United States).[34] Given the multiple registries around the world, compounded by the different priority rules for each jurisdiction, creditors cannot easily determine their priority position with respect to competing creditors and claimants.[35]

Fourth, repossession of a satellite in orbit is virtually impossible and therefore specialized approaches to the enforcement of security interests are required in order to allow the creditor to gain control of the asset in the event of default (such as the cooperation of states where ground stations are located that control satellites).[36]

Finally, a satellite is of limited value to a creditor due to (1) the restrictions on the transfer of satellites under domestic laws,[37] (2) the difficulty of transferring licenses required for the operation of a satellite to a new owner (such as licenses to use the needed orbital positions and radio frequencies

31 See, e.g., Nunziante, *supra* note 29, §20:42 (referencing Article 9 of Regulation (EC) No. 593/2008 of the European Parliament and of the Council which permits member states of the European Union to enforce "mandatory rules" that safeguard critical public interests).

32 See Hazan, *supra* note 17, at 223.

33 See Larsen, *supra* note 21, at 236.

34 Regarding the registration of security interests in the United States see *infra* Section 1.3.

35 See Ospina, S., "The Concepts of Assets and Property: Similarities and Differences, and Their Applicability to Undertakings in Outer Space", in *Proceedings of the Forty-Fifth Colloquium on the Law of Outer Space* 12, 21 & note 17 (2003).

36 See Davis, *supra* note 13, at 459; Hazan, *supra* note 17, at 223.

37 As discussed below, restrictions on the transfer of space assets may arise in the context of export controls. See *infra* Section 2.11.14.

for communications satellites),[38] and (3) the difficulty of repositioning and repurposing a satellite for the use of another owner (since the technical requirements of a satellite may differ for the new owner/operator and the use of on-board fuel for repositioning will reduce the lifespan of the satellite).

Even in this challenging legal environment, practitioners can structure a transaction to minimize the risks posed by existing law. Since the priority of a security interest will likely be governed by the law of the country in which the asset is located and the enforcement of a security interest may depend on the cooperation of the courts where a satellite is located (or where the stations controlling a satellite are located), it is critical that the transaction be structured so that countries with inhospitable laws are avoided. The practitioner must also examine the laws of those jurisdictions where the space asset may be located (or from where it will be controlled) to ensure that any legal requirements for the recognition and enforcement of the security interest have been met under the laws of each jurisdiction. This may require certain contractual elements and the filing of certain documents with state agencies (such as the filing of a financing statement with the appropriate state office under state law in the United States). Any control codes or other information needed to gain control of a satellite when enforcing a security interest should also be placed in escrow in a country that will allow for the easy enforcement of an agreement calling for the release of this information to the creditor upon the default of the debtor. In order to facilitate the transfer of any licenses required for the operation of a satellite to a new owner when enforcing a security interest, the creditor should obtain the consents of the debtor for such transfer in the transactional documents and an assurance that the debtor will take all necessary actions to effect such transfer. Agreements ancillary to the operation of a satellite should also be assigned to the creditor, such as transponder leases, transponder service agreements, or tracking, telemetry and control agreements.

[38] As an example of domestic legal restrictions placed on the transfer of licenses for the operation of satellites, U.S. federal law prohibits the transfer or assignment (including the grant of a security interest) of any license granted by the Federal Communications Commission (FCC) to a satellite operator without the permission of the FCC. 47 U.S.C. §310(d). However, some courts have upheld the creation of security interests in the *proceeds* of an FCC license. See Nesgos, *supra* note 6, §30.06.1.

1.3. *An Example of an Efficient Law of Secured Finance: Article 9 of the U.C.C.*

This section describes the operation of Article 9 of the Uniform Commercial Code (U.C.C.) which has been implemented in all fifty of the United States and is frequently chosen as the applicable law for those satellite finance transactions involving a U.S. party.[39] Article 9 is a creditor-friendly law of secured transactions that facilitates secured lending by allowing for the easy creation of a security interest, clear priority rules, and low-cost enforcement of the security interest. In fact, the Cape Town Convention and its protocols echo many of the provisions found in Article 9, which obviously served as a useful model in the drafting process. However, despite its many positive features, Article 9 is only one of a multiplicity of domestic laws that a court may decide to apply to a dispute arising out of the financing of a space asset that may cross borders prior to being launched or have connections to multiple jurisdictions due to the location and activities of the parties. Even if Article 9 is chosen in the transaction documents, parties cannot be certain that their choice of law will be respected by a court with jurisdiction over the matter.

A security interest can be created under Article 9 of the U.C.C. against any type of space asset including satellites, space stations, launch vehicles, and spacecraft. This interest can be created in favor of a lender upon the mere execution of a security agreement that grants a security interest in the asset.[40] The security interest can encompass the asset in its entirety or be restricted to particular components of the asset, such as the individual transponders of a communications satellite. Security interests can also be created in intangible property, such as launch licenses or licenses to use a particular radio frequency in orbit, as well as other property related to the

---

39 For a more detailed discussion of the provisions of Article 9 of the Uniform Commercial Code see Miller, F.H. (ed.) *Hawkland's Uniform Commercial Code Series* (2012) and Lawrence, L., *Lawrence's Anderson on the Uniform Commercial Code* (2012).

40 The creation of a security interest under Article 9 requires that (1) value be given by the secured party to the debtor (typically in the form of a loan), (2) the debtor has rights in the collateral, and (3) a security agreement describing the collateral be signed by the debtor. The American Law Institute and the National Conference of Commissioners on Uniform State Laws, Uniform Commercial Code §9-203(b) [hereinafter U.C.C.].

space asset, such as transponder leases that the debtor has entered into with third parties.[41]

In order to ensure the priority of its security interest, the secured party will perfect the security interest by filing a financing statement with the Secretary of State's office in that state where the debtor is located. The secured party that is first to file a financing statement (or otherwise perfect a security interest) has priority over competing secured parties.[42] A perfected secured party will also have priority over other competing claimants, such as parties holding a judgment against the debtor and who have converted that judgment into a lien against the debtor's assets (known as a "lien creditor" in the parlance of Article 9).[43] If the debtor is a corporation, the debtor is located (and a financing statement must be filed) in the state of incorporation.[44] When a debtor is a foreign corporation, if a debtor's headquarters are located in a jurisdiction "whose law generally requires information concerning the existence of a nonpossessory security interest to be made generally available in a filing, recording, or registration system as a condition or result of the security interest's obtaining priority over the rights of a lien creditor," the secured party must file in that foreign jurisdiction.[45] If the law of the corporation's jurisdiction does not provide for such a filing system, the debtor will be deemed located in the District of Columbia.[46] In order to avoid any risk arising from the uncertainty as to whether a foreign corporation will be deemed located where it has its headquarters or in the District of Columbia, a secured party would be well

---

[41] A license falls into the catch-all category of "general intangibles" under Article 9. Id. §9-201(a)(42). A transponder lease would be characterized as an "account." Id. §9-201(a)(2). However, see *supra* note 38 regarding restrictions on the transfer of certain licenses.

[42] Id. §9-322(a).

[43] Id. §9-317(a).

[44] Id. §9-307(e).

[45] Id. §9-307(c). To restate this rule, perfection will be governed by the law of a foreign jurisdiction if the headquarters of a foreign corporation is located in a jurisdiction that requires the filing of a document (similar to the filing of a financing statement in the U.S.) in order for the secured party to gain priority over a competing lien creditor – which would include, for example, a party who won a court judgment against the debtor and converted it to a lien against the debtor's assets. Whether or not a foreign jurisdiction qualifies under this section can be a difficult question. A helpful analysis of this question can be found in Rosenberg, Arnold S., "Classification of Foreign Filing Systems", in *Practice Under Revised Article 9* 105 (S. Sepinuck ed., 2008).

[46] U.C.C. §9-307(c).

advised both to perfect its security interest under the law of the foreign jurisdiction and to file a financing statement in the District of Columbia. The state registries containing financing statements are indexed by the name of the debtor and can be searched, frequently online, by the general public. Prospective lenders can therefore quickly discover whether there may be other competing secured parties with priority by searching the U.C.C. registry in that state where the debtor is located.

Potential problems could arise for a secured lender under the U.C.C. if the collateral is located outside of the United States when it comes time to enforce the security interest. The U.C.C. provides that the law of the *jurisdiction in which the collateral is located* governs priority issues.[47] When an asset is located in one of the fifty U.S. states, the secured party will benefit from the clear priority rules of Article 9. However, when an asset is located in a foreign jurisdiction, the priority rules of that foreign jurisdiction will apply. This could result in the secured party's loss of priority depending on the nature of the foreign priority rules. For example, as discussed above, German law grants priority to the security interest that was created first.[48] Although the secured party may have had priority under Article 9's "first to file or perfect" rule, the secured party may have nevertheless created its security interest subsequent to the creation of another security interest in the favor of another secured party – and would therefore lose priority under German law.[49] This issue becomes more complicated when an asset is in orbit since an object in space is located outside the jurisdiction of any country and, as a result, the Article 9 rule applying the law of location of the collateral to priority issues creates uncertainty about the applicable law governing priority.[50] When faced with this situation, a court would have to fall back on a general choice of law analysis. The general U.C.C. choice of law rule states that if "a transaction bears a reasonable relation to [the state

47 Id. §9-301(3)(C).

48 See *supra* Section 1.2.

49 Under the U.C.C. "first to file or perfect" priority rule, Bank A can have priority over Bank B even if Bank B created its security interest first if Bank A filed a financing statement first (a practice known as "prospective filing.").

50 The Outer Space Treaty establishes outer space as an international commons, similar to the high seas, by providing that "[o]uter space, including the Moon and other celestial bodies, is not subject to national appropriation by claim of sovereignty." Treaty on Principles Governing the Activities of States in the Exploration and Use of Outer Space, Including the Moon and Other Celestial Bodies, art. II, Jan. 27, 1967, 610 U.N.T.S. 205.

where the court sits] and also to another state or nation," the parties can choose the law of any of those jurisdictions.[51] If the court finds that the chosen law does not "bear a reasonable relation" to the transaction, the court will apply the U.C.C. in force in the state where the court sits if the transaction "bears an appropriate relation" to that state.[52] Either of these standards will likely be satisfied if one of the parties to the transaction has a connection to the state or if any of the transactional activity occurs within the state at issue. A number of states, such as New York, will apply its law if chosen in the transactional documents even if the transaction has no connection to that state.[53] In practice, the result of this choice of law analysis will not matter provided that the law of one of the fifty United States is determined to apply since Article 9 has been adopted in all states and, as a result, the priority rules in all states are uniform. However, if a court determines that the applicable law regarding priority is the law of a jurisdiction outside the United States, the outcome of a priority question between competing claimants would be settled under that law. To the extent that there is uncertainty under the U.C.C. regarding the applicable law governing priority, this flaw would be remedied by the Cape Town Convention which, if broadly ratified, would provide a single set of rules to govern the financing of space ventures.

In addition to providing clear priority rules, Article 9 allows the secured party to enforce its security interest quickly and at low cost. Upon the default of the debtor, a secured party has the right to repossess the asset and dispose of the asset (by way of sale, lease, license or other manner).[54] The secured party may pursue these remedies without a court order which reduces the expenditure of time and money in connection with enforcing these remedies. The only restriction on the debtor's disposition of the collateral is that the debtor must proceed in a "commercially reasonable" manner.[55]

When the assets are located in the United States or are controlled by parties based in the United States, enforcement will generally proceed without significant obstacles. However, the enforcement of a security interest may

[51] See, e.g., the applicable provisions under the laws of California and New York. Cal. Com. Code §1301; N.Y. U.C.C. Law §1105.

[52] See id.

[53] N.Y. Gen. Obl. Law § 5-1401.

[54] U.C.C. §§9-609(a); 9-610(a).

[55] Id. §9-610(b).

prove problematic if the cooperation of foreign courts or officials is required. For example, if a U.S. bank needs to enforce a security interest in a satellite while the satellite is awaiting launch beside a launchpad at the Baikonur Cosmodrome in Kazakhstan, the bank may need a local court in Kazakhstan to issue an order allowing for an agent of the bank to take possession of the satellite. Whether the Kazakh court will issue such an order is unpredictable. The bank could seek an order from a U.S. court requiring the satellite operator to relinquish possession, but the ability of the court to compel the operator to do so would require the court's personal jurisdiction over the operator and would also depend on the effectiveness of the court's threat of sanctions in compelling the operator to comply with the court order. Personal jurisdiction will likely be found if the operator has been engaging in financial transactions with the U.S. bank. But the operator may nevertheless choose to go ahead with the launch of the satellite in violation of the order of the U.S. court. In that case, the cooperation of the Kazakh courts and local officials would be needed to stop the launch and deliver the satellite to the bank since a U.S. court has no authority to enforce its orders abroad.[56]

An additional feature of Article 9 that serves to protect the interests of the secured party is the ability of the secured party to enforce its security interest even after the debtor has entered insolvency. Article 9 achieves this by giving a secured party priority of the trustee in bankruptcy provided that the secured party is perfected before the debtor enters insolvency.[57] This protection of the secured party's interests is also recognized under U.S. federal bankruptcy law which grants the secured party relief from the automatic stay that is placed on all actions against a bankrupt party's assets and allows the secured party to enforce its security interest against the collateral.[58]

56 The enforcement of U.S. court orders abroad can be problematic. For example, Russian law does not provide for the enforcement of a foreign court order unless a treaty is in place between the Russian Federation and the state where the order was issued. See Morriss, W.D. et al., *Russian Federation* §27.94, in *International Secured Transactions* (Campbell, D., ed. Thomson Reuters 2012).

57 U.C.C. §9-317(a). A trustee in bankruptcy is characterized as a type of "lien creditor" under Article 9 which allows for the application of the general priority rule governing contests between secured parties and lien creditors upon the insolvency of the debtor.

58 Section 362(d) of the Bankruptcy Act allows for the automatic stay to be lifted upon motion by a perfected secured creditor. 11 U.S.C. 362(d).

1.4. *The Cape Town Convention: A New Era of Asset-Backed Finance*

A number of efforts to modernize and harmonize the law of secured transactions have been undertaken by various organizations in the past and these efforts have taken the form of either multilateral treaties or the promulgation of model laws or guidelines.[59] Perhaps the most successful of these efforts has been the Legislative Guide on Secured Transactions which was produced by the United Nations Commission on International Trade Law (UNCITRAL).[60] This guide provides extensive recommendations for states for reforming their laws governing secured transactions and these recommendations largely reflect the approach adopted by Article 9 of the U.C.C.[61] While many states have used this guide when reforming their laws, variations still exist even among those states that have undertaken significant reform and therefore harmonization will not be perfect.

The Cape Town Convention is a significant step forward in the reform of the law of secured transactions because it is an international treaty and therefore promises the highest degree of harmonization among the signatory states. The Convention and the Space Assets Protocol resolve many of the challenges of the secured finance of space assets by creating a uniform international regime to govern asset-backed transactions. Provided that the Protocol is broadly ratified, space companies and their creditors will no longer have to navigate a host of domestic laws of questionable efficacy, but will be able to rely on a single set of rules that govern their transaction. Broad ratification by a significant number of states (or at least of the major space actors) is important because the benefit of the law is in the harmonization of law across borders – and such harmonization can only take place to the extent that states ratify the treaty. However, even if ratification is limited, the benefits of the treaty will be limited to those states that have become a party to the treaty.

---

59 Regarding prior efforts to harmonize the law of secured transactions see McCormack, *supra* note 21, at 102–129; Buxbaum, *supra* note 21, at 328–337; Sundahl, M.J., "The 'Cape Town Approach': A New Method of Making International Law", 44 Colum. J. Transnat'l L. 339, 345–348 (2006); Larsen, P.B., and Heilbock, *supra* note 12, at 714–716.

60 UNCITRAL, Legislative Guide on Secured Transactions (2008), available at www.uncitral.org/pdf/english/texts/security-lg/e/09-82670_Ebook-Guide_09-04-10English.pdf. For a discussion of the UNICTRAL guide see, generally, McCormack, *supra* note 21.

61 Regarding the use of Article 9 of the U.C.C. as a model for the UNICTRAL guide see McCormack, *supra* note 21, at 1.

Among the primary benefits of the Convention is that it solves the problem of the *lex sitae* since the location of the collateral does not affect the application of the Convention to the transaction. The Convention also provides clear and simple rules for the creation of an international interest as well as for determining the priority of competing claimants. At the heart of the system of priority is a centralized registry of international interests that will be available online in order to ensure transparency. This uniformity and clarity is intended to eliminate the uncertainty of the existing patchwork of domestic laws and the increased transactional costs that flow from such uncertainty. In addition, the Protocol provides for creditor remedies that are tailored to the special nature of space assets. As a result, financial institutions will be able to more easily rely on having recourse to a debtor's space assets in the event of a debtor's default. This will enable companies with limited terrestrial assets to seek the benefits of asset-backed finance which will, in turn, reduce the cost of borrowing money – or will even make debt financing available where it would not have been possible prior to the Convention and Protocol.[62] The Protocol will also be beneficial to project finance structures that rely on a strong security package. One commentator has also predicted that the Protocol will have a positive impact on leveraged buyouts in the space industry.[63] The Protocol also applies to the leasing and sale of space assets and promises to provide greater clarity and uniformity in these areas of the law as well.

62 The cost of borrowing money is lowered when banks charge a lower interest rate because they face a lower risk of financial loss. If banks have confidence that they have a first-priority security interest that is easily enforceable, they will not be forced to charge high interest rates to cover potential losses. This effect of reducing the cost of money was concretely illustrated when the Aircraft Protocol to the Cape Town Convention entered into force and the Export-Import Bank of the United States reduced by one-third its exposure fee on the export financing of certain commercial aircraft for buyers in states party to the Convention. Stanford and Poulain, *supra* note 17, at 2. For studies regarding the economic impact of the Cape Town Convention with respect to the aviation industry see Saunders, A., and Walter, I., "Proposed UNIDROIT Convention on International Interests in Mobile Equipment as Applicable to Aircraft Equipment Through the Aircraft Equipment Protocol: Economic Impact Assessment", 23 Air & Space L. 339 (1998); Saunders, A., Srinivasan, A., Walter, I., and Wool, J., "The Economic Implications of International Secured Transactions Law Reform: A Case Study", 20 Univ. of Pennsylvania. J. Int'l Econ. L. 309 (1999).

63 Kozuka, S. and Taniguchi, F., "The Economic Assessment of the Space Assets Protocol to the Cape Town Convention", in *2011 Proceedings of the International Institute of Space Law* 409, 419 (2012).

In addition to containing all of the legal features needed to facilitate asset-backed transactions, the Convention is remarkable in other ways.[64] Perhaps most notably, the structure of the Convention is unique due to its unprecedented use of supplementary protocols. The Convention consists of a base convention that operates in conjunction with the following three protocols:

- Protocol to the Convention on International Interests in Mobile Equipment on Matters Specific to Aircraft Equipment (the "Aircraft Protocol")[65]
- Protocol to the Convention on International Interests in Mobile Equipment on Matters specific to Railway Rolling Stock (the "Rail Protocol")[66]
- Protocol to the Convention on International Interests in Mobile Equipment on Matters Specific to Space Assets (the "Space Assets Protocol")[67]

While the base convention contains the bulk of those fundamental rules that are common to all industries covered by the Convention, the protocols contain rules that are tailored to the needs of a particular industry. This unique structure provided the flexibility needed to respond to the idiosyncratic requirements of the different industries involved in the drafting of the Convention.[68] In addition to its innovative use of protocols, the Cape Town Convention was ground-breaking in other respects as well, such as

---

64 It should be noted that while the Convention does give a creditor an interest in the proceeds of the collateral, the scope of proceeds to which this right attaches is narrow. See *infra* Section 2.12.1.

65 Protocol to the Convention on International Interests in Mobile Equipment on Matters Specific to Aircraft Equipment, art. XXVIII(1), Nov. 16, 2001, S. Treaty Doc. No. 108-10, available at www.unidroit.org/english/conventions/c-main.htm [hereinafter Aircraft Protocol].

66 Protocol to the Convention on International Interests in Mobile Equipment on Matters specific to Railway Rolling Stock, available at www.unidroit.org/english/conventions/mobile-equipment/railprotocol.pdf [hereinafter Rail Protocol].

67 Space Assets Protocol, *supra* note 3.

68 When applying the Convention to a particular transaction, the Convention and the appropriate protocol must be applied together. In some cases the protocol will not only supplement the Convention by addressing issues that are not raised in the Convention, but may also change the rules set forth in the Convention. This need to use two instruments results in a complexity that some commentators have criticized. In order to provide some relief, consolidated texts of the Convention and various protocols have been created. However, these consolidated texts are not authoritative documents, but merely

the degree to which private industry was involved in the drafting process, the primacy placed on commercial expediency, the merging of common law and civil law concepts, and its extensive use of opt-in and opt-out provisions to promote broad ratification.[69]

1.5. *The Genesis and Drafting of the Convention and Space Assets Protocol*

The history of the Cape Town Convention began some twenty-four years before the Space Assets Protocol was opened for signature when the idea of drafting an international convention to cover secured transactions involving high-value mobile equipment was proposed by T.B. Smith QC, the Canadian member of the Unidroit Governing Council, in the wake of the diplomatic conference held in Ottawa for the signing of the Unidroit Convention on International Financial Leasing and the Unidroit Convention on International Factoring.[70] After Unidroit's preliminary studies

---

tools of convenience. As such, the practitioner and scholar should not cite the consolidated texts as legal authority, but should instead work from the Convention and the applicable protocol.

69 Regarding the innovative aspects of the Cape Town Convention see, generally, Gopalan, S., "Harmonization of Commercial Law: Lessons from the Cape Town Convention on International Interests in Mobile Equipment", 9 Law & Bus. Rev. Am. 255 (2003); Sundahl, *supra* note 59. An "opt-out provision" is a provision that automatically applies unless a declaration is made. Official Commentary, *supra* note 4, at 29. In contrast, an "opt-in provision" will only apply if a declaration is made. Id. These features of the convention and protocols allow states the ability to join the Convention and protocols when it would not otherwise have been possible if there had been no possibility of avoiding a problematic provision. For example, a rule providing for the prompt enforcement of a security interest upon the debtor's insolvency conflicts with the policy espoused by certain states to shield the assets of a bankrupt company from grasping creditors. See Wool, J., "The Case for a Commercial Orientation to the Proposed Unidroit Convention as Applied to Aircraft Equipment", 31 Law & Pol'y Int'l Bus. 79, 92 (1999) (explaining that "[f]ew, if any, areas of commercial law are more policy laden than insolvency law."). Because this conflict might prevent many states from ratifying the protocols, each of the three protocols provide that the provision governing the enforcement of a security interest upon insolvency will only apply if a state "opts-in" to the provision by making a declaration. Aircraft Protocol Art. XI(1); Rail Protocol Art. IX(1); Space Assets Protocol Art. XXI(1).

70 Gopalan, *supra* note 69, at 256. For a detailed timeline of the drafting of the Convention and the three protocols through the adoption of the Convention and Aircraft Protocol see Official Commentary, *supra* note 4, at 531–539.

predicted that considerable economic benefits would flow from a new convention of this nature, Unidroit formed a Study Group to create the initial draft of a convention that would both modernize and harmonize the law of secured transactions with respect to a narrow band of high-value collateral for which there existed a pressing need for such a convention.[71] These categories of collateral were eventually reduced to aircraft, rolling stock, and space assets.[72]

When assembling the Study Group, Unidroit pursued a highly progressive approach by granting private industry a strong voice in the drafting process. The aviation industry in particular had a high profile role in the process through the participation of the International Air Transport Association (IATA), an industry group representing approximately 265 airlines, and an Aviation Working Group, which was composed of representatives from the aviation manufacturing and finance sectors and was co-chaired by Boeing and Airbus. In the beginning of the process, the Aviation Working Group and the IATA worked with the representatives of the space, railway and other industries in an attempt to craft a single convention that would serve the needs of all the industries.[73] This first phase of the drafting process continued for eight years until, in 1996, the process stalled due to the difficulty of creating a single set of rules that would meet the different needs of the different industries.[74]

It was at this juncture that the creation of the Cape Town Convention took an interesting turn. Frustrated by the delay in the drafting process, the Aircraft Working Group asked for the IATA's assistance in modifying the process to allow the aircraft industry to move forward independently.[75] In response to Boeing's request, the General Counsel to the IATA, Lorne Clark, proposed that each of the three industries involved in the drafting of the

[71] UNIDROIT Study LXXII - Doc.2, at 7 (1989).

[72] The working group had initially contemplated applying the Convention to a broader collection of assets including airframes, aircraft engines, helicopters, oil rigs, containers, railway rolling stock, registered ships and space assets. Stanford 1999, *supra* note 4, at 244.

[73] Clark, *supra* note 4, at 3.

[74] Id. at 4–5; Cuming 2000, *supra* note 4, at 1094–95. One issue over which the three industries held different opinions was the nature of the registry where international interests would be registered in order to establish the priority of the creditor's claim. UNIDROIT News, 2 Unif. L. Rev. 308, 310 (1997).

[75] Clark, *supra* note 4, at 5; Stanford 1999, *supra* note 4, at 244.

convention proceed independently with the creation of an industry-specific protocol that would supplement a base convention.[76] This proposal, which would allow industries the flexibility to develop at their own pace rules that were responsive to the needs of their particular industry, was adopted by the Unidroit Study Group during its third session held in Rome in January 1997.[77]

Following the adoption of the protocol approach, Unidroit reorganized the drafting process: Unidroit was to oversee the drafting of the base convention, while an Aircraft Protocol Group was formed to draft the Protocol to the Convention on International Interests in Mobile Equipment on Matters Specific to Aircraft Equipment.[78] The Aircraft Protocol Group consisted of the Aircraft Working Group, the IATA, and the International Civil Aviation Authority.[79] A Space Working Group was organized by Peter Nesgos, Scott Siegel, and Dan Schechter to draft a protocol specific to space assets.[80] Similarly, a Rail Working Group was organized by Howard Rosen to draft a protocol for rolling stock.[81] Other working groups addressed specific issues, such as registration, public international law, and insolvency.[82]

The protocol approach succeeded in "breaking the logjam" and allowing the convention to be completed with respect to the first category of assets. On November 16, 2001, the Convention and the Aircraft Protocol were complete and opened for signature at the diplomatic conference convened for this purpose in Cape Town. The Convention entered into force with respect to aircraft on March 1, 2006, when the Aircraft Protocol entered into force.[83]

---

76 Clark, *supra* note 4, at 5.

77 UNIDROIT Study LXXII - Doc. 32, at 1–2 (Dec. 1996); UNIDROIT News, 2 Unif. L. Rev. 308, 308, 312(1997).

78 Clark, *supra* note 4, at 5.

79 Official Commentary, *supra* note 4, at 6; Clark, *supra* note 4, at 5.

80 Official Commentary, *supra* note 4, at 534.

81 Id.

82 Id. at 536.

83 The third instrument of ratification required for the Cape Town Convention's entry into force was deposited by Nigeria on December 16, 2003. See *Status of the Convention on International Interests in Mobile Equipment*, available at www.unidroit.org/english/implement/i-2001-convention.pdf. However, the Convention will only enter into force with respect to a particular category of assets when the protocol for that category of assets comes into force. See *infra* Section 2.3.5. The Aircraft Protocol entered into force on March 1, 2006. See *Status of the Protocol to the Convention on International Interests in*

The Rail Protocol was opened for signature in Luxembourg on February 23, 2007, while the Space Assets Protocol was opened for signature on March 9, 2012 in Berlin. Of the three protocols, the Space Assets Protocol was the last to be completed due to the complexity of transactions involving space assets and the special concerns of the space industry. A preliminary draft Protocol was completed by a Working Group composed of representatives from a broad swath of space actors including manufacturers, financiers, and operators of space assets, as well as the United Nations Office for Outer Space Affairs, the European Space Agency, the European Centre for Space Law, and the International Bar Association. This draft protocol was approved by the Unidroit Governing Council in 2001. It would take another eleven years to create a final version of the protocol.[84]

As of the time of writing, the Rail Protocol has not yet received the four ratifications required for its entry into force. Although signed by five states and the European Union, it has only been ratified by Luxembourg.[85] One year after being opened for signature, the Space Assets Protocol has been signed by four states (Burkina Faso, Germany, Saudi Arabia, and Zimbabwe).[86] Germany signed the protocol on November 21, 2012, while the other three states signed the protocol at the Diplomatic Conference.[87] No instruments of ratification have yet been deposited – and ten are required for the protocol to enter into force. If Germany's signature is a first sign of European support for the protocol, ten ratifications may not be far off. However, as discussed in the following section, the willingness of states to ratify the protocol has been hampered by the sharp criticism leveled by members of the commercial space industry in the course of its drafting.

Mobile Equipment on Matters Specific to Aircraft Equipment, available at www.unidroit.org/english/implement/i-2001-aircraftprotocol.pdf.

84 A detailed chronology of the working group meetings and procedural steps that led to the completion of the Protocol has been published by Unidroit online and is available at www.unidroit.org/english/workprogramme/study072/spaceprotocol/conference/background.htm.

85 See Status of the Luxembourg Protocol to the Convention on International Interests in Mobile Equipment on Matters specific to Railway Rolling Stock available at www.unidroit.org/english/implement/i-2007-railprotocol.pdf.

86 See Status of the Protocol to the Convention on International Interests in Mobile Equipment on Matters specific to Space Assets, available at www.unidroit.org/english/implement/i-2012-spaceassets.pdf.

87 Id.

1.6. *Industry Reception of the Space Assets Protocol*

Although the clarity and international harmonization provided by the Cape Town Convention and its protocols would seem to provide undeniable benefits, space industry organizations have vigorously opposed the adoption of the Protocol. The general concern of industry is that the existing models of space finance are sufficient and that the Protocol "adds an unnecessary supra-national layer of law" to the financing of space assets.[88] Additional specific concerns raised in the course of the drafting process by industry groups, including the Satellite Industry Association, include the following:[89]

- The vagueness of the definition of "space assets"
- The ability of parties with a security interest in a component to prevent the enforcement of a security interest against the principal asset
- The stay on remedies against space assets used for public services
- The priority of insurers' salvage rights with respect to secured lenders
- The criteria for identification of space assets for registration
- The assignment of debtor's rights
- The need for existing owners of satellite to register their interest

---

88 Statement of the Satellite Industry Association on the Revised Preliminary Draft Protocol to the Convention on International Interests in Mobile Equipment on Matters Specific to Space Assets (18 October 2010), available at www.esoa.net/upload/files/news/unidroit/20101018sia.pdf [hereinafter the SIA 2010 Statement].

89 Global Satellite Industry Denounces Unidroit Protocol (Brussels, 9 March 2012), available at www.esoa.net/upload/files/news/20120309_PR_UNIDROIT.pdf; SIA 2010 Statement, *supra* note 88, at 2–4; Continuing Issues of Concern Regarding the Unidroit Draft Space Protocol to the Cape Town Convention on International Interests in Mobile Equipment dated 16 April 2010, available at www.esoa.net/upload/files/news/unidroit/20100416industry.pdf. A startup satellite communications company, Avanti Communications, joined the SIA in expressing reservations about the Protocol. ISPL Workshop on Unidroit Space Protocol, A New Satellite Operator Perspective, available at www.esoa.net/upload/files/news/unidroit/satoperator.pdf. However, the objections expressed by Avanti are rather vague in nature by simply stating that existing law is sufficient and the benefits of the Protocol are few while it imposes additional transactional burdens. See also Sa'id Mosteshar, "Financing Space Assets: The Unidroit Solution Examined," available at www.esoa.net/upload/files/news/unidroit/analysisoflegalissues.pdf; Aganaba, T., "Assessing the Support for the Space Assets Protocol to the UNIDROIT Cape Town Convention", Report of International Institute of Space Commerce (2009), available at www.iisc.im/documents/IISC_-_09_-_P.12.1%5B2%5D_1.pdf.

In light of these concerns, the United States delegation to the Diplomatic Conference in Berlin recommended that the Protocol not be adopted until further study of its economic benefit could be undertaken.[90]

The concerns expressed by the space industry were taken into account in the course of finalizing the text of the Protocol and many of the issues were resolved in the final text of the Protocol. For example, the priority of insurers' salvage rights with respect to competing secured lenders was left to existing domestic law by a clear statement in the Protocol that salvage interests would not be affected by the Convention.[91] The concern about the need for existing owners to register in order to protect their interests was also resolved by the drafters' decision to exempt pre-existing interests from the operation of the Convention.[92] The final version of the Protocol also placed multiple limitations on the stay on remedies against space assets used for public services in order to strike the proper balance between the public interest and the needs of creditors. A more detailed discussion of the evolution of the Protocol's approach to each of these issues can be found in the appropriate sections of Chapter 2.

The debate about whether the Space Assets Protocol will ultimately benefit the space industry will only be settled with the passage of time. However, the weight of experience and economic theory indicates that a harmonized, transparent, and creditor-friendly law of secured transactions will facilitate the financing of space ventures.[93] The fact that the drafters resolved a number of the concerns raised by the space industry in the final version of the Protocol also suggests that the industry's fears should dissipate. Although the implementation of the Protocol will entail some initial costs as the industry accustoms itself to a new legal regime, the new law promises to be an improved system that will ultimately bring greater certainty and efficiency to space transactions.

---

[90] Comments submitted by Governments, Organisations and Representatives of the International Commercial Space, Financial and Insurance Communities, UNIDROIT 2012, DCME-SP–Doc. 6 Add. 1 (February 2012), available at www.unidroit.org/english/conventions/mobile-equipment/conference2012/dcme-sp-06-1-e.pdf.

[91] See *infra* Section 2.12.2.

[92] See *infra* Section 2.3.3.

[93] For an eloquent and comprehensive explanation of the benefits of security interests see McCormack, *supra* note 20, at 1–38 (stating, for example, that "[s]ecured credit is variously said to be the oil of the economy and the engine of economic growth" and that "[c]ommonly banks argue for stronger, broader and more effective security rights."). There are, of course, some academics who challenge the economic benefit and fairness of secured credit. See, e.g., Hudson, J., "The Case against Secured Lending", International Review of Law and Economics (1995).

Chapter Two

The Operation of the Convention and Space Assets Protocol

With the first chapter having set the stage with an explanation of the various methods of financing space ventures and the challenges of doing so in the current legal environment, this chapter turns to the technical operation of the Cape Town Convention and the Space Assets Protocol. The Convention and Protocol together contain the features of an efficient law of asset-backed finance that promise to facilitate the financing of space ventures. However, the full benefit of this new international treaty system will only be realized when the two instruments are ratified by a significant number of states, including the major space-faring countries. Only then will the flawed patchwork of domestic laws be replaced by a harmonized set of efficient rules governing these transactions.[1]

The mechanics of the Convention and Protocol can be understood as involving five basic parts: (1) the definition of the types of interests and transactions that are governed by the Convention, (2) the technical requirements for creating these interests (3) the remedies available to the parties to the covered transactions, (4) the priority rules that determine the rights of parties that are simultaneously asserting their rights with respect to the same asset, and (5) the mechanism for a party to ensure its priority through the registration of its interest. Of course, these basic issues are complicated by a number of subsidiary issues and considerations. This chapter explains the operation of the Convention and Protocol in a manner intended to provide a clear and unified description of the complex provisions contained in the two instruments. Generous citations to the relevant provisions are provided in the course of the chapter and the operation of the

[1] It should be noted, however, that even upon broad ratification of the Protocol, certain issues are not addressed by the Convention and Protocol and, as a result, recourse to other applicable domestic law will still be required with respect to resolving those issues (such as, to give one example, the basic question of the validity of a contract).

provisions is, in many cases, illustrated through the means of hypothetical scenarios. Recommendations for the practitioner are also offered along the way in order to provide guidance in the application of the Convention to finance transactions.

2.1. *The Nature of an International Interest*

The Cape Town Convention creates a new type of legal interest called an "international interest."[2] Although this term is newly coined, it is a composite of three long-standing types of interests: a security interest, the interest held by the conditional seller under a title reservation agreement, and the interest of a lessor.[3] The natures of these interests are distinct, but they share features that justify their treatment in the same legal regime as "international interests." In addition to their common purpose of financing the acquisition of space assets (either through a true sale with a security interest, a conditional sale, or a lease), each of these transactions requires remedies that will enable the seller, chargee, or lessor to retrieve or resell the equipment in the event of default. In other words, each of these transactions provides for recourse to the underlying asset – thus allowing all of these transactions to be characterized as "asset-backed" transactions.

A variety of terms are used to identify parties to the transactions governed by the Convention. In the context of a security interest, the party granting the security interest is the "chargor", while the party being granted the security interest is the "chargee."[4] The parties to a title reservation agreement are referred to as the "conditional seller" and the "conditional buyer."[5] Finally, the parties to a lease transaction are the "lessor" (who grants the right to use the equipment) and the "lessee" (who makes payments in return for the right to use the equipment).[6] The Convention also uses the two catch-all terms "creditor" and "debtor."[7] The term "creditor" includes chargees, conditional sellers, and lessors, while the term

[2] Convention on International Interests in Mobile Equipment, arts. 1(o), 2(2), Nov. 16, 2001, S. Treaty Doc. No. 108–10, available at www.unidroit.org/english/conventions/c-main.htm [hereinafter Cape Town Convention].

[3] Id. Art. 2(2).

[4] Id. Art. 1(ii).

[5] Id. Art. 1(e) & (f).

[6] Id. Art. 1(q).

[7] Id. Art. 1(i) & (j).

"debtor" includes chargors, conditional buyers, and lessees. When a provision in the Convention or Protocol addresses an issue that is peculiar to a specific type of transaction, then the specific term is used. For example, when the remedies available to a chargee under a security agreement are described in Article 8 of the Convention, the term "chargee" is used in order to make clear that these remedies do not apply to conditional sellers or lessors (which parties have their separate remedies spelled out in Article 9). In other cases, a provision may apply to all three types of transactions, in which case the general terms "creditor" and "debtor" are used (as is seen in Article 11 of the Convention which explains the meaning of default – an issue that is common to all transactions).

2.1.1. *Security Interests*

Briefly stated, a security interest arises when a debtor (the chargor) takes a loan from a creditor (the chargee) – or otherwise incurs a monetary obligation to the creditor by, for example, purchasing an item on credit – and then grants the creditor recourse to the debtor's property in the event that the debtor defaults on the repayment of the obligation. The debtor's property serves as collateral for the obligation, thus providing assurance to the creditor that even in the event of default the obligation will be repaid (since the creditor will have the right to sell the collateral or otherwise realize upon its value). The nature of a security interest is captured in the Convention's definition of "security agreement":[8]

> an agreement by which a chargor grants or agrees to grant to a chargee an interest (including an ownership interest) in or over an object to secure the performance of any existing or future obligation of the chargor or a third person.

The debtor can also grant a security interest in its property to secure an obligation owed by a third party to the creditor.[9] The concept of a security interest is referred to in a variety of ways in different legal systems, including terms such as security interest, charge, lien, pledge, hypothecation, and chattel mortgage. In order to avoid confusion, the generic term "security interest" was chosen by the drafters as a catch-all phrase, following the

[8] Id. Art. 1(ii). The type of obligation that can be secured on collateral is not limited to monetary obligations, but can include other obligations of the debtor.

[9] Id. Art. 1(ii).

model of Article 9 of the Uniform Commercial Code which uses the same term. The definition of "security interest" and "security agreement" in the Convention displays an artful blending of Uniform Commercial Code terminology (security interest) and civilian terminology (chargor/chargee) that indicates how the concept of this transaction was intended to transcend any one legal tradition and be truly international in nature.

2.1.2. *Conditional Sales*

A conditional sale is closely related to a security interest that arises in the context of a credit sale. However, whereas a seller who takes a security interest in collateral does not have title to the collateral, a conditional seller retains title in the sold item until the final payment is made (and other conditions of the sale are fulfilled).[10] In the event of default, the conditional seller takes back the item that was sold and is thus protected from financial loss. A conditional sale is documented by a title reservation agreement.[11] A security interest has greater range in that the creditor can take a security interest in any of the debtor's property, while a conditional seller will only have recourse to the sold item. The remedies available to the conditional seller will also necessarily differ from those enjoyed by the secured party.[12]

2.1.3. *Leases*

Leases were also included in the Convention because the lease is a common method employed by airlines to acquire the use of aircraft – as well as being a common way for telecommunications companies to acquire the use of satellite transponders.[13] The basic nature of a lease is familiar to most. The lessor provides the use of its equipment for a certain period of time (the "term" of the lease) in exchange for periodic payments.[14] Title remains

---

10 Id. Art. 1(ll).

11 "Title reservation agreement" is defined in the Convention as "an agreement for the sale of an object on terms that ownership does not pass until fulfilment of the condition or conditions stated in the agreement." Id.

12 See *infra* Section 2.11.5.

13 For a description of satellite leasing from a practical perspective see Nesgos, P., "Satellites and Transponders", in Wong, J. and Dubin, B. (eds.) *Equipment Leasing* (2012).

14 The nature of a lease is described in the convention's definition of "lease agreement" which is defined as "an agreement by which one person (the lessor) grants a right to possession or control of an object (with or without an option to purchase) to another person (the lessee) in return for a rental or other payment." Cape Town Convention, Art. 1(q).

with the lessor and, at the end of the term of the lease, the lessee has no further right to use the equipment and returns it to the control of the lessor. It is this right to regain control of the equipment, among other things, that is provided for in the Convention.

2.1.4. *Resolving Questions about the Nature of a Transaction*

In some cases, there may be a question as to whether a particular transaction should be properly characterized as a security agreement, a title reservation agreement, or a leasing agreement. These are distinct and mutually exclusive types of transactions. Neither a lease transaction nor a conditional sale involves a security interest. Nor can a security interest be deemed a lease or conditional sale. The nature of a particular transaction (whether it involves a security interest, lease, or conditional sale) will affect the application of certain provisions of the Convention and Protocol. For example, the types of remedies available will differ depending on the nature of the transaction. The determination of a transaction's nature may be obvious in most circumstances, but in some cases the nature of the transaction may be debatable. Questions regarding the nature of a transaction must be resolved by other applicable law (which will likely be the applicable domestic law).[15] For example, a transaction that is nominally structured as a lease may be more accurately described, and more properly treated, as an installment sale with a security interest retained by the "lessor" who is, in fact, the seller. A court may make such a determination if, for example, the lease cannot be terminated by the lessee and at the end of the term the lessee becomes owner of the asset without having to make any further payments.[16] In substance, this arrangement is no different than an installment sale. Similarly, under certain domestic laws a title retention agreement may be treated as a sale with a security interest (as is the case under the Uniform Commercial Code in the United States).[17] The applicable law is to be determined under the "rules of private international law of the forum state," which forum can be selected by agreement of the

15 Id. Art. 2(4).

16 The Uniform Commercial Code contains a detailed provision to clarify when a so-called "lease" is actually a sale with a security interest. The American Law Institute and the National Conference of Commissioners on Uniform State Laws, Uniform Commercial Code §1-203 (2012) [hereinafter U.C.C.].

17 Id. §1-201(b)(35).

parties.[18] The Convention also makes clear that a transaction that creates a security interest cannot also qualify as a conditional sale or a lease.[19]

2.1.5. *The Requirement of a "Uniquely Identifiable Space Asset"*

In order to qualify as an "international interest" that is subject to the Convention and the Space Assets Protocol not only must the nature of the interest be one of the three types just mentioned, but the interest must also involve a space asset that is "uniquely identifiable."[20] The meaning of "uniquely identifiable" as well as the definition of "space asset" are addressed in the following section.

2.2. *The Definition of "Space Asset"*

As explained above, the Space Assets Protocol extends the scope of the Cape Town Convention to "space assets" (bringing into effect what was originally contemplated in Article 2 of the Convention).[21] Although the term "space asset" is mentioned in the Convention, it is only defined in the Protocol.[22] Pursuant to this rather complicated definition, not every object in space can be treated as a space asset subject to the Convention. Instead, there are a number of requirements that must be met before a transaction involving the object will be governed by the Convention.[23]

18 Cape Town Convention, Arts. 5(3), 42(1).

19 Id. Art. 2(2).

20 Id. Arts. 2(2) & 2(3)(c); Protocol to the Convention on International Interests in Mobile Equipment on Matters Specific to Space Assets, in *Final Act of the Diplomatic Conference for the Adoption of the Draft Protocol to the Convention on International Interests in Mobile Equipment on Matters Specific to Space Assets*, art. I(2)(k).UNIDROIT Doc., DCME-SP–Doc. 43 (9 March 2012), available at www.unidroit.org/english/workprogramme/study072/spaceprotocol/conference/documents/dcme-sp-43-e.pdf [hereinafter Space Assets Protocol].

21 Cape Town Convention, Art. 2(3)(c).

22 Space Assets Protocol, Art. I(2)(k).

23 For a discussion of the early evolution of the term "space asset" in the Convention see Ospina, S., "The Concepts of Assets and Property: Similarities and Differences, and Their Applicability to Undertakings in Outer Space", in *Proceedings of the Forty-Fifth Colloquium on the Law of Outer Space* 12 (2003); Ribbelink, O.M., "The Protocol on Matters Specific to Space Assets", 12 European Rev. Private L. 37, 39 (2004) (explaining that the Protocol replaced the initial term "space property" with "space assets" in order to avoid confusion

First, the object must be man-made. An asteroid or minerals mined from the Moon or another celestial body will not qualify as a space asset (although at some point the transformation of mined metal through human effort into a new object would allow for the application of the Convention to that object, assuming that the other features of a space asset are met).

Second, the object must be "uniquely identifiable." This is a requirement under Article 2 of the Convention as well as under the Protocol.[24] What is required in order to be uniquely identifiable is not further clarified and so this element should be satisfied by any feature that enables identification of the specific asset – whether through serial number, by physical features, or in some other manner. This ability to identify the specific asset is essential to the registration of the space asset in the International Registry (as discussed in greater detail below).[25] It should be noted that a specific description of the asset is not required in a security agreement, conditional sales agreement, or lease to create an international interest under the Protocol – although the asset must still be uniquely identifiable.[26]

Third, the object must either "be in space or designed to be launched into space." This allows for the creation of an international interest not only in an object that is already in space, but also for a space object that has not yet been launched. The broad nature of this aspect of the definition captures payloads and spacecraft (or components thereof) in any stage of their existence whether in a state of manufacture, in storage, en route to a launch site, or on the launch pad (since at any one of these stages the object has been designed to be launched into space).[27]

The original definition of what constitutes a "space asset" required that an asset must be either "in space or *intended* to be launched into space." Some industry participants considered this to be an impossibly vague criterion since it would be difficult to ascertain the existence of such intentions (or to document exactly when such intentions arose). In order to address

---

with the term "space object" that is used in the space treaties and to more accurately reflect the commercial finance nature of the Protocol); Yun Z., "Revisiting Selected Issues in the Draft Protocol to the Cape Town Convention on Matters Specific to Space Assets", 76 Journal of Air Law and Commerce 805, 814–815 (2011).

24 Cape Town Convention, Art. 2(2); Space Assets Protocol I(2)(k).

25 See *infra* Section 2.5.2.

26 See *infra* Section 2.4.3.

27 It should be noted that an international interest continues after a space asset returns from space to Earth. Space Assets Protocol, Art. III(c).

this concern, the drafters changed "*intended* to be launched into space" to "*designed* to be launched into space." Although the question of whether or not an item has been designed for use in space may be debatable in some cases, the special design requirements of an asset that will survive the space environment (such as ruggedization) make this criterion a substantial improvement over the more nebulous question of intention.

The phrase "in space or designed to be launched into space" also raises the question of the delimitation of space – or, in other words, what is "space". The Protocol provides a separate definition of "space", but avoids drawing any bright lines by simply defining "space" as "outer space", thus leaving it to other law to determine precisely where space begins.[28] Although, as discussed in greater detail below, the issue of where space begins has been hotly debated for many years by academics, it is unlikely to be a concern for financiers of satellites since satellites are unarguably designed to be launched into orbital space.[29] On the other hand, someone could in good faith raise the question whether a suborbital space plane that reaches a maximum altitude of 105 kilometers (just above the von Kármán line which lies 100 kilometers above the Earth and is widely viewed as the limit of space) is a space asset. The same question could be raised with respect to a space plane that reaches an altitude of 95 kilometers – just short of the von Kármán line. In connection with this issue, it should be noted that the Protocol will only apply to a vehicle that is "primarily designed for use in space" and does not apply to a vehicle, such as a suborbital spaceplane, that is designed primarily to travel in air space and is only in space for a brief time.[30] However, the question of the delimitation of space is still relevant if, for example, a space plane was in low-altitude space for an extended period of time, as is being contemplated for the next generation of space planes that will be capable of point-to-point travel.

Fourth, in order to come within the scope of the Convention the asset must be a "spacecraft", a "payload", or a part of a spacecraft or payload. What comprises a spacecraft is left to broad interpretation – as is the definition of "payload". Illustrative examples of spacecraft and payloads are provided, but they are non-exhaustive. The definition of spacecraft explicitly includes any "satellite, space station, space module, space capsule,

---

28 Id. Art. I(2)(j).

29 See *infra* Section 3.3 regarding the definition of space.

30 See *infra* Section 2.3.2.

space vehicle, or reusable launch vehicle," which should capture any satellite, space station and vehicle (whether manned or unmanned) that is currently being developed or imagined by the space industry. The additional type of space asset – a payload – could be said to overlap to some degree with the concept of a spacecraft in that a satellite could be deemed to be both a spacecraft and the payload of the rocket on which it was launched.[31] The term "payload" could also be interpreted more narrowly as referring to that portion of a satellite that is separate from the satellite bus (in other words, the equipment that carries out the specific function of the satellite, such as transponders for communications or cameras for remote sensing). The inclusion of payloads as a type of space asset clarifies that any equipment being sent into space as payload, regardless of its nature and purpose, can be the subject of an international interest separate from any international interest that may attach to the launch vehicle carrying that payload (provided that the other requirements of the definition of space asset are met).

Fifth, the asset must be separately registrable pursuant to the regulations governing the registration of international interests in space assets.[32] These regulations are issued by the Supervisory Authority pursuant to Article XXIX of the Protocol. This requirement for separate registration of each spacecraft, payload, or part thereof is necessitated by the asset-specific nature of the registration system.[33] Whether an asset is separately registrable will depend on whether the asset will be uniquely identifiable under the requirements set forth in the registration regulations. The Supervisory Authority may determine that a certain part or payload cannot be uniquely identified – in which case it will not be capable of separate registration and, therefore, will not qualify as a "space asset."[34]

If these five requirements are met, an international interest can attach to the asset in question. Moreover, the provisions of the Convention and Protocol will extend to all parts or equipment "installed, incorporated or attached" to the space asset as well as to "all data, manuals and records

31 The definition of space asset takes a broad view of what constitutes a payload by stating that any payload would come within the definition of space asset "whether telecommunications, navigation, observation, scientific or otherwise." Space Assets Protocol, Art. I.2(l).

32 Id. Arts. I(2)(k)(ii) & (iii).

33 For further discussion regarding the nature of the registry see *infra* Section 2.5.1.

34 See *infra* Section 2.5.2 regarding the mechanics of registration.

relating thereto."[35] In some cases, a separate international interest may be created in a specific part or component of a space asset (provided that the above requirements are met with respect to that component) – but the component can also be captured by the international interest in the principal space asset of which it is a part. The extension of the concept of space asset to embrace all data, manuals and records relating to the asset expands the definition beyond the equipment so that it includes information, in written or electronic form, that pertains to the operation of the asset. This is necessitated by the sophisticated nature of satellites and other space assets and the need for control codes and other information required to operate the equipment.

2.3. *The Sphere of Application of the Convention*

The following sections explain the circumstances under which the Convention will apply to a particular transaction. Understanding when the Convention applies is critical so that the parties and the practitioner take the Convention and Protocol into account when structuring the transaction and drafting the transactional documents. A court handling a dispute arising from a transaction involving a space asset will also need to understand when the Convention governs the dispute. In addition to an explication of the general provisions regarding the application of the Convention, the following sections also discuss the application of the Convention to purely domestic or "internal" transactions, the application of the Convention to interests in existence prior to the entry into force of the Convention with respect to space assets, and the requirements for the entry into force of the Convention as to space assets.

2.3.1. *General Rules Regarding the Sphere of Application of the Convention*

The Convention applies to those transactions that create an international interest when, at the moment the transaction is closed, the chargor, conditional buyer, or lessee is "situated" in a state that is a party to the Convention and Protocol.[36] The Convention also applies to sales of space assets when

[35] Space Assets Protocol, Art. I(2)(k).

[36] Cape Town Convention, Arts. 1(j), 3(1) & 49.

the seller of a space asset is situated in a state party to the Convention and Protocol.[37] A debtor (or seller) will be deemed to be "situated" in a state if any of the following conditions is met:[38]

- It is incorporated or formed under the law of the state;
- It has its registered office or statutory seat in the state;
- It has its center of administration in the state;
- It has its principal place of business in the state; or
- It has its habitual residence in the state (if it has no place of business).

This test for the applicability of the Convention is drafted to capture as many transactions as possible by providing several ways of coming within the scope of the Convention. This will serve the goal of harmonizing the law by expanding the application of the Convention to any transaction that can meet one these requirements.

2.3.2. *The Application of the Convention to Vehicles Only Temporarily in Space*

In some cases, there may be some question whether a particular vehicle should be subject to provisions of the Space Assets Protocol or the Aircraft Protocol. This may be the case for high-altitude airplanes or spacecraft that fly both in air space and in low earth orbit. In order to avoid this complexity, the Protocol establishes that the application of the appropriate protocol will hinge on whether the vehicle is "primarily designed for use in space."[39] If designed predominantly for use in space, the asset will be governed by the Space Assets Protocol, even when the asset is not in space.[40] Conversely, the Protocol will not apply to transactions involving a vehicle that is primarily designed for flight in air space. This raises the question of whether the Space Assets Protocol applies to transactions involving the suborbital space planes of the type being developed by Virgin Galactic which will briefly enter outer space, but soon thereafter drop back into air space. The Protocol answers this question in the negative by explicitly stating that the "Protocol does not apply to an aircraft object merely because it

37 Space Assets Protocol, Art. IV(1).
38 Cape Town Convention, Art. 4.
39 Space Assets Protocol, Art. II(3).
40 Id.

is designed to be temporarily in space."[41] However, space planes that spend a significant amount of time in space (as was true for the Space Shuttle) would be subject to the Protocol.

Whether a transaction governing a particular space plane that is excluded from the Space Assets Protocol would be subject to the Aircraft Protocol must be analyzed under the provisions of the Aircraft Protocol.[42] If a transaction is governed by the Aircraft Protocol, the practitioner must comply with the requirements of the Aircraft Protocol with respect to the creation and registration of the international interest which differ in certain important respects from the mechanics of the Space Assets Protocol. Other differences between the Aircraft Protocol and the Space Assets Protocol must also be taken into account, a description of which is beyond the scope of this study.

Merely because a transaction involving a vehicle is not subject to the Space Assets Protocol does not mean that the activities involving that vehicle are not subject to space law in general. It could well be that the transaction involving a suborbital space plane would be governed by the Aircraft Protocol and be subject to space law as well. If this is the case, there will be intersections between the Aircraft Protocol and space law of which the practitioner should be aware. While these intersections may be similar to those described herein with respect to the Space Assets Protocol, the effect of these intersections may be quite different in certain respects when the Aircraft Protocol is involved because the Aircraft Protocol contains no provision similar to Article XXXV of the Space Assets Protocol giving supremacy to the space treaties over the Aircraft Protocol. As a result, the space treaties may be trumped by conflicting provisions of the Aircraft Protocol under the doctrine of *lex posterior*. Those instances in which a different result is reached with respect to the Aircraft Protocol's interaction with space law are discussed in those sections below concerning export

---

[41] Id. Art. II(4).

[42] With respect to aircraft powered by a jet engine, the Aircraft Protocol will only apply to transactions involving an aircraft with (1) an airframe built to carry at least eight persons or 2750 kilograms of goods and (2) an engine with at least 1750 lbs of thrust. Protocol to the Convention on International Interests in Mobile Equipment on Matters Specific to Aircraft Equipment, arts. I(2)(a); I(2)(b); I(2)(e); and II(1), Nov. 16, 2001, S. Treaty Doc. No. 108–10, available at www.unidroit.org/english/conventions/c-main.htm [hereinafter Aircraft Protocol].

controls, jurisdiction, and the application of the Rescue and Return Agreement.[43]

2.3.3. *The Application of the Convention to Pre-existing Interests*

An early draft of the Protocol required holders of pre-existing interests in satellites to register their interests to ensure their priority over interests registered after the protocol entered into force. This provision caused concern among industry representatives because it would have not only created an administrative burden for every satellite owner and creditor with a security interest in a satellite, but also raised the specter of a loss of priority if such registration did not take place. In order to ease the transition to the new law and protect the priority of existing creditors, Article XL(2) was inserted into the Protocol. This provision exempts any pre-existing interests from the operation of the Protocol and explicitly states that a pre-existing interest "retains the priority it enjoyed" under other applicable law.[44] Subjecting such pre-existing interests to the new rules of the Convention would unfairly prejudice the holders of such interests who would have relied on the priority provided laws under the applicable law when the interests arose.

For purposes of determining whether the Convention applies to an existing interest, the "effective date" of the Convention is the later of (1) the date the Convention enters into force with respect to space assets and (2) the date when the state in which the debtor is situated when the interest was created becomes a signatory state to the Convention and Protocol.[45] The one exception to this exemption of pre-existing interests arises in limited cases involving space assets subject to separate international interests that are physically linked.[46]

43 See *infra* Section 2.11.14 (regarding export controls), Section 3.6.2 (regarding jurisdiction), and Section 3.5.3 (regarding the Rescue and Return Agreement).

44 Space Assets Protocol, Art XL(2).

45 Id. Art. XL(3)(a). A debtor is "situated" in a state if it (1) has its center of administration in the state, (2) has its place of business in the state (if there is no center of administration), (3) has its principal place of business in the state (if there is more than one place of business), or (4) has its habitual residence in the state (if the debtor has no place of business). Id. Art. XL(3)(b).

46 Id. Art. XL(2). See *infra* Section 2.11.13 regarding the enforcement of remedies involving assets that are physically linked.

2.3.4. *Internal Transactions*

Upon ratification or accession to the Convention, a state may make a declaration that the Convention (and hence the Protocol) will not apply to "internal transactions."[47] A transaction qualifies as "internal" if (1) all parties to the transaction as well as the relevant space asset are located – at the time of the transaction – in the state that has made the declaration and (2) the interest arising from the transaction, such as a security interest, is registered on a domestic registry in that state.[48] For the purpose of qualifying as an "internal transaction," a space asset is deemed to be "located" in a state not only when the asset is physically located in the territory of the state on Earth, but also when the asset is in space if the asset has been registered by the state under the Outer Space Treaty, the Registration Convention, or the 1961 UN Resolution for International Cooperation in the Peaceful Uses of Outer Space (General Assembly Resolution 1721 (XVI) B of 20 December 1961).[49]

Despite a state's declaration to exclude internal transactions from the scope of the Convention, the declaration is actually of limited effect. Even if a declaration has been made, Article 8(4) (regarding the requirement to give notice prior to the sale or lease of a space asset), Article 9(1) (regarding the ability to vest ownership of a space asset in the chargee), Article 16 (regarding the establishment of a registry), Chapter V (regarding the mechanics of registration), Article 29 (regarding priority), and "any provisions of th[e] Convention relating to registered interests" will still apply to an internal transaction.[50] The Convention does not clarify what these other "provisions relating to registered interests" are. However, the drafter of the Convention has clarified in one of his writings that these would include, for example, Article 30 (international interests in the

47 Cape Town Convention, Art. 50(1).

48 Id. Art. 1(n).

49 Space Assets Protocol, Art I(3).

50 Cape Town Convention, Art. 50(2). The continued application of Article 8(4) to internal transactions is somewhat troublesome because the notice prior to sale or lease required under this provision contemplates that the sale or lease is taking place pursuant to Article 8(1) of the Convention – which does not apply to internal transactions. Roy Goode himself characterizes this as a "slip." Goode, R., "Cape Town Convention on International Interests in Mobile Equipment: A Driving Force for International Asset-Based Financing", 7 Unif. L. Rev./Revue de droit uniforme 3 (2002).

context of insolvency), Article 35(1) (priority of assignments), and Article 39 (priority of non-consensual rights or interests).[51]

Of particular importance is that, while the remedies contained in the Convention will not apply to internal transactions, the priority provisions of the Convention will apply to internal transactions even if a declaration was made excluding such transactions from the Convention's scope. For example, if a state with a domestic law requiring a court order prior to the disposition of collateral has made such declaration, a court order would be required prior to the creditor selling the collateral – but the question of whether the creditor had priority over competing claimants would still be resolved by the rules under the Convention. This means that a party holding an interest that arises from an internal transaction must still ensure its priority by entering a registration on the international registry. However, the party would not register an international interest, but would instead register a "notice of a national interest" – a "national interest" being an interest that arises from an internal transaction.[52] The priority of that interest would then be determined by the date of registration. As is true for a party that registers an international interest, a party that has registered a notice of national interest must discharge the registration "without undue delay" upon written demand of the debtor when the underlying obligation has been extinguished.[53]

2.3.5. *Entry into Force*

The Convention will only enter into force with respect to space assets when the Space Assets Protocol has entered into force.[54] There would be no benefit in having the Convention enter into force without the Protocol also entering in force since the Convention cannot operate on its own and, by its terms, only applies to those assets covered by a protocol.[55] Thus, the Convention will enter into force with respect to space assets when the

---

51 Id.

52 Cape Town Convention, Arts. 1(r), 1(t) & 20.

53 Id. Art. 25.

54 Id. Art. 49(1). The third instrument of ratification required for the Cape Town Convention's entry into force was deposited on November 16, 2001. See *Status of the Convention on International Interests in Mobile Equipment*, available at http://www.unidroit.org/english/implement/i-2001-convention.pdf.

55 Cape Town Convention, Art. 49(1).

Space Assets Protocol enters into force, which will occur on the later of (1) first day of the month after the expiration of three months after the tenth state ratifies the Protocol and (2) the date of the deposit of a certificate by the Supervisory Authority confirming that the International Registry is operational.[56] Transactions involving space assets will only be subject to the Convention with respect to a particular state if that state is party to both the Convention and the Protocol. After the Protocol enters into force, if a state ratifies, accepts, approves or accedes to the Protocol, the Protocol will enter into force with respect to that state three months after the deposit of the instrument of ratification, acceptance, approval, or accession.[57]

2.4. *Creating an International Interest*

The Convention and Protocol take a minimalist approach to the formation of an international interest. The formalities for the creation of an international interest are few and these few formalities are themselves flexible. This allows parties to easily grant an international interest while minimizing the risk that the international interest will be invalidated due to the failure to adhere to transactional formalities. There are three basic requirements for the creation of an international interest: (1) an agreement in writing, (2) the power of the debtor to dispose of the asset, and (3) proper identification of the asset in the agreement. A fourth requirement is added in the case of security interests. Each of these requirements is more fully discussed in the following paragraphs.

2.4.1. *The Requirement of a Written Agreement*

The first requirement for the creation of an international interest is that the parties to the underlying transaction must have an agreement "in writing" (whether a security agreement, a title reservation agreement, or a leasing agreement).[58] This requirement is further elucidated in the definition of "writing" which requires:[59]

---

56 Id.; Space Assets Protocol, Art. XXXVIII(1).
57 Id. Art. XXXVIII(2).
58 Cape Town Convention, Arts. 1(a) & 7(a) with chapeau.
59 Id. Art. 1(nn).

> a record of information (including information communicated by teletransmission) which is in tangible or other form and is capable of being reproduced in tangible form on a subsequent occasion and which indicates by reasonable means a person's approval of the record.

This definition acts to further reduce the formalities of creating an international interest by allowing not only agreements written on paper, but also permitting the agreement to be on any medium provided that the agreement (1) can be reproduced in a tangible form (for example, printed out on paper) and (2) gives some indication of the party's approval (without restricting such indication to the traditional signature in ink). Under this definition, an email sent by a chargor granting a security interest to the chargee at the end of which the chargor types his name should be sufficient (although the chargor should also indicate his title and the name of his company if the writing is intended to bind the company). That only the chargor is required to sign the security agreement is not made perfectly clear, but it is not negated in the Convention either. In the absence of such clarity, it is a matter of other applicable (likely domestic) law. If the Uniform Commercial Code governed, for example, a security agreement signed only by the debtor would be sufficient.[60]

2.4.2. *The Requirement of the "Power to Dispose"*

The second requirement for the creation of an international interest in a particular asset is that the chargor, conditional seller, or lessor must have the power to dispose of the asset.[61] This is a fundamental concept that does not permit a person to grant an interest in an asset unless the person owns the asset (or otherwise has the "power to dispose" of the asset). It is not necessary that the debtor enjoy outright ownership of the asset to satisfy this requirement.[62] Such a "power to dispose" might include the ability of an agent to grant an interest by means of his apparent authority to do so (even if actual authority is lacking), the ability of a party without ownership rights to transfer good title to a *bona fide* purchaser, or the mere right to use

60 U.C.C. §9-203(b)(3)(A).

61 Cape Town Convention, Art. 7(b).

62 Compare a similar concept in the Uniform Commercial Code, a debtor can create a security interest if the debtor either has "rights in the collateral or the power to transfer rights in the collateral." U.C.C. §9-203(b)(2).

an asset under a lease – which right can then be transferred to another party through a sublease.[63]

2.4.3. *The Requirement of the Sufficient Identification of the Asset*

The third requirement for the creation of an international interest is that the asset must be "identified in conformity with the Protocol" in the parties' agreement.[64] In contrast to the Aircraft Protocol, the Space Assets Protocol takes a relatively relaxed approach to the description of space assets in a security agreement, title retention agreement, and lease agreement. While the Aircraft Protocol requires the specific description of the asset by serial number, name of manufacturer, and model designation, the Space Assets Protocol (like the Rolling Stock Protocol) allows for a more generic and open-ended description of the assets that are subject to the agreement.[65] Under Article VII of the Space Assets Protocol, the description of the space asset in the agreement is sufficient if any of the following three options are satisfied:[66]

- the asset is described by item;
- the asset is described by type; or
- the agreement states that it covers all present and future space assets.

These three options run from the more specific to the least specific description.[67] A description by "item" is the most specific of the three options and contemplates the description of a particular asset, although the manner in which the asset is identified is not mandated.

---

63 For an example of a domestic law allowing a bona fide purchaser for value to acquire good title from a seller with voidable title see U.C.C. §2-403. See also Goode, R., *Convention on International Interests in Mobile Equipment and Protocol Thereto on Matters Specific to Aircraft Equipment: Official Commentary* 176 (2nd ed. 2008) [hereinafter Official Commentary].

64 Cape Town Convention, Art. 7(c).

65 Aircraft Protocol, Art. VII; Rolling Stock Protocol, Art. V(1); Space Assets Protocol, Art. VII(1).

66 Space Assets Protocol, Art. VII(1).

67 The Protocol's approach to the description of assets also reflects an apparent influence of the Uniform Commercial Code on the drafting of the Protocol since the Uniform Commercial Code provides a "safe harbor" for sufficient description of collateral by stating that a description will be sufficient if the collateral is identified specifically, by category, or by a type of collateral defined in the Uniform Commercial Code. U.C.C. §9108(2).

The second option is to describe more generally the "type" of asset that is subject to the agreement, which would allow an international interest to attach to any number of assets held by the grantor, seller or lessor of the type described. For example, a security agreement granting a security interest in "all satellites" owned by the debtor would be sufficient to give rise to an international interest in any satellites owned by the debtor. This description by type reduces transactional costs by eliminating the need to individually identify each asset that is subject to the agreement – and also eliminates the risk of a mistake being made in the specific description of a particular asset (which error could result in the invalidation of the international interest).

The third option casts the widest net by allowing a debtor to grant a security interest in all its space assets, whether such assets are currently in existence or will be acquired by the debtor after the conclusion of the security agreement. This allows for the creation of a "floating lien" that will attach to subsequently acquired space assets at the time that the debtor acquires rights in the assets. This option shares the benefits of the description by "type" mentioned above, but also eliminates the need to enter into multiple subsequent agreements in order to create an international interest in subsequently acquired assets, thus reducing transactional costs even further. Of course, an international interest in subsequently acquired assets would only technically arise when the debtor actually acquired rights in the assets (and in acquiring such rights would also acquire the necessary "power to dispose" of the asset) since this is a separate requirement that must be fulfilled in order for an international interest to be created.[68] The Protocol also allows for a variation on this third option by permitting a debtor, seller or lessor to carve out particular assets or types of assets from the scope of the agreement.[69] For example, a debtor could grant a security interest in "all existing and future space assets owned by the debtor with the exception of all remote sensing satellites." However, the language of the Protocol appears to allow only for the carving out of assets from a general statement that the agreement covers "all present and future space assets," but does not allow simply for the creation of an international interest in all existing and future assets of a particular type. This is an absurd result that could not have been intended by the drafters, but practitioners should be

68 Space Assets Protocol, Art. VII(2).

69 Id. Art. VII(1)(d).

aware of the possibility that a description of future assets of a particular type may be challenged as insufficient under the black letter law of the Protocol.

2.4.4. *The Requirement of a Statement of Obligations Secured and the Permissibility of Future Advances*

In the case of a security interest, the Convention imposes the additional requirement that the security agreement "enables the secured obligations to be determined, but without the need to state a sum or maximum sum secured."[70] The security agreement need not within its four corners set forth the sums owed. The agreement need only describe the nature of the obligations so that the parties are able to determine what obligations are secured under the agreement. This description can extend to obligation of all types as well as future obligations, thus permitting blanket language such as "all existing and future obligations owed by the debtor to the creditor."[71] The Official Commentary to the Convention explains that the primary purpose of this provision was to make clear that there was no requirement under the Convention to state any precise figure nor even to state a maximum obligation that could be secured under the agreement (as is required under some domestic laws).[72] This again reduces the number of technical formalities when creating an international interest.

Under certain progressive secured transactions laws, a security agreement can not only secure the original loan issued by the creditor, but can also act to secure all additional loans, or "future advances," made by the creditor to the debtor. This is true, for example, under the Uniform Commercial Code in the United States provided that the security agreement explicitly allows for securing such future advances.[73] Other legal systems place limitations on the ability to secure future advances. In the most extreme example, under one interpretation of classical Islamic law security interests will not automatically secure future advances, which means that the security agreement must be amended to explicitly cover a new loan every time a new advance is made.[74] More progressive laws not only allow

70 Cape Town Convention, Art.7(d).

71 Official Commentary, *supra* note 63, at 177.

72 Id.

73 U.C.C. §9-204(c).

74 Sundahl, M., "Iraq, Secured Transactions and the Promise of Islamic Law", 40 Vanderbilt Journal of Transnational Law 1301, 1328 (2007).

for future advances to be covered automatically by the original security agreement, but also give the same priority rights to the creditor when enforcing the security interest with respect to these future advances.[75] However, limitations are often placed on this right, as is the case under the Uniform Commercial Code, which allows a third-party buyer of collateral to take free of a security interest to the extent that it covers a future advance that was made either 45 days after the sale or was made with the creditor's knowledge of the sale.[76]

The Convention takes a progressive approach to the issue of future advances. Under the Convention, all future advances can be secured by the original security agreement, provided that the agreement is clear that all such future advances are covered.[77] Moreover, the creditor will enjoy the same priority rights with respect to future advances, since no exception is made regarding future advances under the general priority rules. Nor are any other limitations placed on the creditor's rights regarding future advances. For example, a third party buyer of collateral takes the collateral subject to any prior registered security interests, regardless of whether the security interest secures only an initial loan or future advances.[78]

2.5. *Registering an International Interest*

The successful creation of an international interest gives the secured party, the conditional seller, and the lessor the right to avail themselves of the remedies set forth in the Convention and Protocol in the event of default. It is not necessary to register the international interest in order to take advantage of these remedies. However, it is necessary to register the international interest on the International Registry created by the Cape Town Convention (and maintained by the appointed Registrar) in order to ensure the priority of the secured party, conditional seller, or lessor over other competing claimants to the space asset.[79] The question of priority will always be a critical issue since there will undoubtedly be competing claimants in the

---

75 U.C.C. §§ 9-322(a), 9-323(a) Comm. 3.

76 Id. §9-323(d).

77 Cape Town Convention, Art. 7(d) (stating that the security agreement need only "enable the secured obligations to be determined," but need not state a "maximum sum secured").

78 Id. Art. 29(3)(a).

79 See *infra* Section 2.6 regarding the priority rules applying to competing claimants.

event that a debtor reaches the point of financial exigency that triggers a default. These competing claimants who will be seeking to attach the assets of the debtor can take multiple forms. Some may be lenders with security interests in the same collateral. Others may be parties who hold court judgments of various types (whether in the nature of commercial disputes or tort actions) in their favor and are attaching assets in satisfaction of the judgments. Another type of party with a competing interest in the collateral could be a buyer who purchased the collateral from the debtor (despite the collateral being subject to an international interest) and who wishes to maintain ownership over the collateral. The centralized online registry created by the Convention, together with the clear priority rules described below, allows for the transparency with respect to the order of priority among such competing claimants that is essential for an efficient and reliable system of conducting the transactions covered by the Convention and Protocol. Banks are more likely to lend money for the purchase of space assets if they are guaranteed a first-priority security interest in the collateral. Providing such assurance is easier under the Convention because (1) priority is assured under the clear rule that the first to register has priority over competing claimants to the asset and (2) banks can easily search online to determine whether any other creditor has registered an international interest relating to a specific asset. If no such registration is found, the bank can easily register their international interest through the online system and thus establish its first-priority right to foreclose on the asset in the event of default. This section describes the nature of the registry and the mechanics of registration.

2.5.1. *The Registry*

The Convention and Protocol call for the appointment of a Supervisory Authority to create an International Registry that will contain the registrations and recordings of the covered interests and then appoint (and supervise) a Registrar to manage the registry.[80] During the Diplomatic Conference for the adoption of the Protocol, the International Telecommunications Union was invited to serve as the Supervisory Authority.[81] At the time of writing, the Supervisory Authority had yet to be

[80] Cape Town Convention, Arts. 16 & 17; Space Assets Protocol, Art. XXVIII.

[81] Resolution Relating to the Establishment of the Supervisory Authority of the International Registry for Space Assets, in *Final Act of the Diplomatic Conference for the Adoption of*

identified. In addition to being a single, centralized registry that will eliminate the need for parties to search multiple domestic registries, the registry is accessible online at all times so that parties can receive instantaneous information about existing registrations 24 hours a day.[82] This allows transactions to move forward with greater efficiency than would be possible if multiple domestic registries had to be searched and search results would only be available by less efficient means, for example, by mail. The Supervisory Authority also promulgates the regulations that govern the operations of the registry, including the processes for registering interests and searching the registry.[83]

2.5.2. *Mechanics of Registration*

Either party to a transaction may register an international interest.[84] However, the consent of the other party must be received prior to registration. In the absence of such consent, the registration will not be valid.[85] Such consent must be in writing and would normally be given in the security agreement, sales agreement, or lease agreement (although consent to the filing of a prospective international interest would be given in a separate document).[86] The consent must be submitted to the Registrar electronically prior to registration.[87] However, the Registrar has no duty to confirm the authenticity of the consent.[88] Amendments to a registration can also be made by either party subject to the same mechanics, including the consent requirement, applicable to registrations.[89]

In addition to the consent of the other party, the only requirement under the Convention and Protocol for a valid registration is that the space asset

---

the Draft Protocol to the Convention on International Interests in Mobile Equipment on Matters Specific to Space Assets, UNIDROIT Doc., DCME-SP–Doc. 43, Annex III (9 March 2012), available at www.unidroit.org/english/workprogramme/study072/spaceprotocol/conference/documents/dcme-sp-43-e.pdf [hereinafter Space Assets Protocol].

82 Space Assets Protocol, Art. XXXII(5).

83 Cape Town Convention, Art. 17(2)(d).

84 Id. Art. 20(1).

85 Id. Art. 19(1).

86 Regarding the broad meaning of "in writing" see *supra* Section 2.4.1.

87 Cape Town Convention, Art. 18.1(a).

88 Id. Art. 20(2).

89 Id. Arts. 16(3) & 20(1).

be sufficiently identified.[90] Because the types of space assets subject to a separately registrable international interest will likely include a broad variety of equipment (given that even parts of a spacecraft or payload could qualify as a space object), the Protocol leaves the identification requirements for registration to the registration regulations.[91] The possible methods for identifying a particular part will vary depending on the type of part. In some cases, the part may carry a serial number and the Supervisory Authority may decide to require the serial number in such case. Other methods of identification may be required in other cases. The Supervisory Authority will make this determination in light of the particular nature of the asset. And if a particular part does not allow for identification under the regulations, it will not qualify as a space asset and cannot be made subject to an international interest.

A specific description of the asset is necessary because the registry is searchable by asset (as explained below) rather than by the name of the debtor, as is the case with the state registries under Article 9 of the Uniform Commercial Code.[92] An asset-specific registry makes it easier for a prospective lender, buyer, or lessee to determine whether another party has a registered interest in a particular asset because they need only search for the asset and will be able to see immediately whether any interests have been registered against the asset. The disadvantage of an asset-based registry is that it requires that each asset be described with specificity, which poses the risk that an error may be made in the description, such as in the transcription of the serial number, and that this error will render the registration invalid resulting in the secured party's loss of priority.

In light of the fact that the registry of international interests will be indexed by asset, it is critical that assets be identified in a consistent manner to ensure that all registrations against a particular asset are discovered upon a search in the registry for that asset. The initial approach to the description of space assets for purposes of registration required different descriptors to describe a space asset when the interest was registered prior to launch (which included the name of manufacturer, serial number, and model number) than the descriptors to be used to describe a space asset when the interest was registered post-launch (which included the date of launch, launch site, and launch provider). Critics of this approach feared

---

90 Id. Art. 18(1); Space Assets Protocol, Art. XXX.

91 Space Assets Protocol, Art. XXX.

92 U.C.C. §9-519(c).

that this bifurcated approach to describing assets would result in confusion that might place a creditor's priority in jeopardy. Responding to this concern, the drafters abandoned this approach. Although the method of describing a space asset under the final version of the protocol will vary depending on the type of asset (since assets of different types will have different identifying marks or features), the method of describing a particular type of asset will be consistent and will no longer change depending on whether the interest was registered pre-launch or post-launch.

A registration is valid upon entry of the required registration information by the Registrar into the International Registry and the registration is made "searchable."[93] A particular registration is "searchable" when (1) the International Registry has assigned to it a file number, (2) the registration information is stored in "durable form", (3) the registration is made publicly accessible through the registry, and (4) the asset is sufficiently described in the registration (as discussed above).[94] It is this critical moment of achieving a valid registration that determines the priority of the international interest. Once the registration is transmitted to the Registrar, the Registrar is required to enter the registration into a searchable database in the order of receipt pursuant to Article 18.4 of the Convention.[95] This requirement ensures, in the event that Party A transmits a registration to the Registrar before Party B transmits a registration against the same asset, that the Registrar will not give priority to Party B by entering its information into the registry first. If the Registrar does not comply with Article 18.4 and enters registrations out of order, the Registrar is liable for compensatory damages arising from the error.[96] In order to provide some protection against this scenario, the Convention requires the registration file to contain the date and time of the receipt of the registration.[97]

The Convention and Protocol allow for a Contracting State to designate one or more "entry points" for submitting a registration to the International Registry.[98] For example, a state may wish to use its domestic registry of security interests as the entry point to the Registry. Under the registration

---

93 Cape Town Convention, Art. 19.

94 Id. Arts. 19(3) & 19(6); Space Assets Protocol, Arts. XXX & XXI(2); see also Official Commentary, *supra* note 63, at 343.

95 Cape Town Convention, Art. 18(4).

96 Id. Art. 28(1).

97 Id. Art. 18.4.

98 Id. Art. 18.5; Space Assets Protocol, Art. XXXI.

regulations for aircraft, a state may designate "authorizing entry points" or "direct entry points."[99] When a registration is submitted through the authorizing entry point, authorization can then be given for the information to be further transmitted to the International Registry. In contrast, a registration submitted to a direct entry point can be automatically submitted to the International Registry. The entry points cannot be used to register a national interest or a registrable non-consensual right or interest that arise under the laws of another state.[100]

2.5.3. *Registration of Prospective International Interests*

The Convention permits creditors to secure the priority of their international interests before all of the elements required to create an international interest are in place. This can be done by registration of a "prospective international interest."[101] The ability to register an international interest prospectively is a valuable and necessary feature of the Convention's mechanism that allows for financial institutions to ensure their priority at the very outset of a financing. Without the ability to register prospectively, a lender would only be able to register after the debtor had acquired the asset subject to the international interest – and by this time the bank would likely have disbursed funds to the debtor to fund the purchase of the asset. This would leave the bank in the dangerous position of having disbursed funds before assuring that it had a first-priority security interest in the collateral. In order to avoid this, the Convention allows for the registration of prospective international interests. For example, a creditor may enter into a security agreement that provides for the creation of a security interest in a particular satellite when the debtor acquires ownership of the satellite.[102] The creditor could register this "prospective" international interest before the final elements of an international interest, such as the debtor's power to dispose of the asset, fell into place, or – perhaps more

99 Official Commentary, *supra* note 63, at 436 (Regulations and Procedures for the International Registry §12).

100 Space Assets Protocol, Art. XXXI.

101 Cape Town Convention, Arts. 1(y) & 16(1)(a).

102 Practitioners should note that the definition of "prospective international interest" requires prospective interest is intended to ripen into a full international interest "*upon the occurrence of a stated event*", which would seem to require an agreement describing the event (such as acquisition of ownership by the debtor) upon which a full international interest will arise. *Id.* Art. 1(y).

importantly – before the creditor released any funds to the debtor. This gives the creditor assurance of its priority before it places its funds at risk.

The benefit of registering prospectively is that the priority of the international interest will be determined according to the date of registration of the *prospective* international interest (provided that the registration of the prospective interest is still valid when the interest arises).[103] Registration of a prospective interest is only possible, given the aspect-specific indexing of the registry, if the asset subject to the prospective interest can be described in the same manner required for registration of a true international interest.[104] The Convention does not require the creditor to amend the registration when the prospective international interest ripens into a true international interest as long as the information in the registry meets the requirements for a valid registration of an international interest (otherwise the registration must be amended to provide any missing information information).[105] As is the case for true international interests, either party can register a prospective international interest provided that the other party has given written consent.[106]

2.5.4. *Duration and Discharge of a Registration*

Once entered on the registry, a registration will remain effective indefinitely, unless it is discharged or if an expiration date is stated in the registration.[107] If an expiration date is stated in the registration, the registration will be "discharged" (in other words, will no longer be effective) as of that date. A discharged registration will not be removed from the registry. A notice of discharge will be attached to the registration and this file will be maintained on the registry as a historical record.[108] Either party can change the expiration date and extend the effectiveness of a registration provided that the registration has not yet expired and the other party has given its consent to the change in writing.[109]

---

103 Id. Art. 19(4).

104 Space Assets Protocol, Art. XXX.

105 Cape Town Convention, Art. 18(3).

106 Id. Arts. 18.1(a) & 20(1).

107 Cape Town Convention, Art. 21.

108 Official Commentary, *supra* note 63, at 214.

109 Cape Town Convention, Art. 20(1).

A registration can be discharged at any time by the party "in whose favour it was made", that is, the party who holds the interest that is registered.[110] This would include a secured party, conditional seller, or lessor whose international interest is registered.[111] An assignee of an international interest also has the right to discharge the registration of the assignment. In some cases, there may be multiple parties who benefit from a registration, as is the case when an international interest is assigned. The written consent of the assignee must also be obtained before the holder of an international interest discharges an international interest that was transferred by assignment. Similarly, the registration of an international interest that has been assigned as security to an assignee benefits both the holder of the international interest and the assignee (since the value of the assignee's collateral – the secured loan – depends to a large extent on the assurance that the international interest securing the loan has first priority, a question that depends on the continued registration of the international interest). In this case, the written consent of the assignee must be obtained. For example, if Bank A assigns as security its loan (along with the international interest securing the loan) to Bank B, Bank B must give written consent before Bank A discharges the international interest. In both of the foregoing cases, the priority of the assignee hinges on the continued validity of the original registration of the international interest.[112]

The secured party is required to discharge the registration if the underlying obligation secured by the space asset has been satisfied.[113] For example, if a satellite operator has paid off its loan obligation secured on its satellites to Bank X, Bank X must discharge the international interest so that the satellite operator is free to take out a new line of credit against its satellites from another bank (which bank will then be able to be assured of priority when it sees that no other international interests are registered against the satellites). A secured party must discharge the registration "without undue delay" following the receipt (or delivery) of a written demand by the debtor sent to the secured party's address stated in the registration.[114]

---

110 Id. Art. 20(3).

111 The debtor, buyer or lessee may also discharge the registration, but only if the other party has given its written consent. Id. Art. 20(3).

112 Official Commentary, *supra* note 63, at 208.

113 Cape Town Convention, Art. 25(1). This obligation to discharge upon the satisfaction of the underlying obligation also applies with respect to a non-consensual right or interest that has been registered.

114 Id. Art. 25(1).

A conditional seller has the same obligation to discharge a registration when the conditions of title transfer have been met by the buyer.[115] With respect to the registration of a prospective international interest (or a prospective assignment of an international interest), the secured party, conditional seller, or lessor (or assignor) must discharge the registration "without undue delay" upon the mere receipt (or delivery) of a written demand by the other party provided that no "value" has been given, that is, that no funds have been disbursed or assets transferred to the debtor, and no contractual commitment to give value exists.[116] At this point in the transaction, the secured party, seller or lessor is at no risk and is not harmed by the discharge of the registration. The Protocol clarifies the time frame for discharging the registration under these circumstances by requiring that the holder of the prospective international interest (or prospective assignee) "take such steps as are within its power to procure the discharge of the registration no later than *ten calendar days* after the receipt of the demand."[117] Of course, if a registration has improperly been made (if, for example, it has been registered against the wrong party or without the debtor's consent) or if the registration contains errors, the party in whose favor it was made must discharge the registration or correct the errors through amendment "without undue delay" following the receipt of a written demand from the debtor.[118]

2.5.5. *Searching the Registry*

The ability of a prospective creditor, buyer, or lessee to easily search the registry of international interests is critical for the efficient operation of the Convention. Through the easily accessible, transparent, and reliable registry of international interests financiers can have the confidence necessary to disburse the funding that is required for space ventures. Any person or entity wishing to determine whether any registrations have been made against a space asset can search the registry online.[119] If the registry for international interest under the Aircraft Protocol serves as an indication of what the Space Assets Registry will look like, we should expect an online

---

115 Id.

116 Id. Art. 25(2).

117 Space Assets Protocol, Art. XXXII(3) (emphasis added).

118 Cape Town Convention, Art. 25(4).

119 Id. Art. 22(1).

registry that is available day and night any day of the year.[120] The registry will be searchable by the same criteria required under the registration regulations for the proper description of a space asset to achieve a valid registration.[121] For example, a prospective lender who is considering issuing a loan secured on a type of space asset that must be described by serial number when registering an international interest in the asset can search for any existing registrations against the asset by entering the serial number of the asset into the search engine.[122] When a person submits a search request to the Registrar, the Registrar will send that person a search certificate electronically which will either (1) state any information submitted in prior registrations against the asset, together with the date and time of such registrations or (2) state that there are no registrations against the asset.[123] If prior registrations are reported, then the prospective lender will have been put on notice that there *may* be a valid interest that would have priority over any security interest that the prospective lender plans to register. However, the discovery of a prior registration does not tell the prospective lender everything he needs to know. The lender does not know the amount of the underlying obligation secured by the existing security interest, whether the security interest is merely prospective, or whether the security interest is even valid.[124] These details must be discovered through conversations with the debtor (who should be forthcoming with the information if he wishes to secure the loan). If, on the other hand, a search of the registry turns up no prior registrations against the asset, the prospective lender can be almost certain that it will be able to acquire a first-priority security interest against the asset upon registration. There is a slight risk at this point if another party has submitted a registration to the registrar, but the information has not yet been entered into the registry. The time lag between submission and entry should be minimal, but it should be kept in mind by the practitioner, who should advise the client not to disburse funds until the new lender's registration appears on the registry without any prior registrations.

---

120 The registry for international interests in aircraft can be found online at www.aviareto.aero.

121 Id. Art. 19(6); Space Assets Protocol, Arts. XXX & XXXII(2).

122 Space Assets Protocol, Art. XXX.

123 Cape Town Convention, Art. 22(2). From an evidentiary standpoint, the certificate issued by the Registrar serves as *prima facie* (but rebuttable) proof of the facts stated in the certificate regarding prior registrations. Id. Art. 24.

124 Id. Art. 22(3).

2.5.6. *Errors in Search Results*

Given the critical importance of the reliability of the registry, errors made in the course of registration (or which work their way into the registry in other ways) can have catastrophic repercussions. Hundreds of millions of dollars can be disbursed in reliance on the information provided by the registry and an error in this information can result in equivalent losses. In the event that a search of the registry returns information containing errors that cause damage to a party, the Registrar can be held liable if the errors were caused by the Registrar's employees or a malfunction of the registration system.[125] Such damage could be cause, for example, if a prospective lender conducts a search and the results indicate that no international interests have been registered against the asset that is to be financed as a result of the Registrar's failure to properly index a prior registration against the same asset. The lender may then proceed with the transaction under the assumption that it will have a first-priority security interest, which could result in serious losses when it turns out that the party that registered first has a senior security interest. There are, however, limits to the Registrar's liability. First, no liability will be found if the malfunction was the result of an "inevitable and irresistible nature" that could not have been avoided by the best practices of electronic registration (assuming that the Registrar was following such best practices).[126] Second, the Registrar is not liable for errors that were originally transmitted to the Registrar by the party undertaking the registration.[127] In contrast, the Supervisory Authority enjoys immunity from any such claims to the extent immunity is granted under the rules that govern the entity.[128]

2.6. *Priority*

The extent to which a law of secured transactions provides the assurance needed by financiers to bring down the cost of financing depends largely

125 Id. Art. 28(1). The courts in the jurisdiction where the administrative center of the Registrar is located has exclusive jurisdiction over liability claims against the Registrar. Id. Art. 44(1).

126 Id. Art. 28(1).

127 Id. Art. 28(2).

128 Id. Art. 27(2); Space Assets Protocol, Art. XXVIII(2).

on the clarity of its priority rules. If a bank cannot be assured a first-priority security interest in the debtor's collateral, that bank will either be unwilling to lend money or will do so at a high rate of interest in order to protect itself against losses following a default. The Cape Town Convention excels in this respect by adopting a simple "first to file" priority rule. Under this rule, priority is determined simply by the date of registration, which results in a priority system that is predictable and not easily challenged since the question of priority can be reduced to the simple question of who filed first.

2.6.1. *The "First to File" Rule*

The "first to file" rule is expressed with concise elegance in Article 29 of the Convention:[129]

> A registered interest has priority over any other interest subsequently registered and over an unregistered interest.

The first component of the "first to file" rule is that a registered international interest has priority over an unregistered interest.[130] This rule gives a registered interest priority over any type of unregistered interest including, for example, a lien creditor that has converted an unpaid judgment arising from a tort claim to a lien against the asset or an unsecured creditor (such as a parts supplier that sold parts to the debtor on open account, that is, on unsecured credit).[131] Therefore, holders of an international interest should register their interests in order to defeat the claims of any party with an unregistered interest. The second component of the rule addresses the situation where two holders of an international interest, both with registered interests in the same asset, are competing for the right to pursue remedies with respect to the asset. For example, two banks with a security interest in the same satellite may both want to seize control of and sell the satellite upon the default of the debtor. In this scenario, the secured party who registered first has priority and will have the right to recoup its losses first from the proceeds of the sale.[132] These rules apply strictly without

---

[129] Cape Town Convention, Art. 29(1).

[130] Cape Town Convention, Art. 29(1).

[131] Id. Art. 1(mm). Note, however, that some unregistered interests may still have priority over a registered international interest if a state has made a declaration to that effect under Article 39 of the Convention. See *infra* Section 2.12.4.

[132] Id. Art. 29(1).

regard to whether the prevailing party had knowledge of the other security interest.[133] For example, Party A may have been granted a security interest in a satellite before Party B, but if Party B registers its security interest first, even though it is fully aware of Party A's interest, Party B will have priority. Moreover, Party B will continue to have priority with respect to enforcing the security interest to recoup funds that it disbursed after Party B acquired knowledge of Party A's security interest.[134] The application of the priority rules does not look into the minds of the parties, but is instead applied mechanically with an eye only to the date of registration. The Convention does not resolve priority contests between unregistered interests (which provides another motivation to register). Such contests must be resolved in accordance with applicable domestic law.[135]

Conditional buyers and lessees of a space asset may find themselves in the situation where another party holds an international interest in the same asset. For example, a bank may have a registered security interest in a satellite that is later leased to the lessee. If the lessor defaults on its obligations to the bank, the bank has the right to enforce its security interest against the satellite (perhaps by selling it). The question then arises whether the lessee will still be able to use the satellite pursuant to its lease. The Convention states that, in such a situation, the lessee does not have the right to continue to use the satellite. Under Article 29 of the Convention, the interest of a conditional buyer or lessee is subject to an interest registered prior to the registration of the interest held by the conditional seller or lessor.[136] Conversely, the buyer or lessee takes free of any interest that was not registered earlier (even if the buyer or lessee had actual knowledge of that interest before buying or leasing the asset).[137] In another scenario, a lessee may enter into a sub-lease with a sub-lessee. In this case, the sub-lessee would be subject to the lease interest of the head lessor registered

---

133 Id. Art. 29.2(a).

134 Id. Art. 29.2(b).

135 For example, under the Uniform Commercial Code the priority would be accorded to the party that was the first to file a financing statement with the appropriate Secretary of State's office or was otherwise the first to "perfect" a security interest under the Uniform Commercial Code. U.C.C. §9-322(a)(1). If neither security interest is perfected, priority goes to the first security interest to attach, i.e. the security interest that arose first. Id. §9-322(a)(3).

136 Cape Town Convention, Art. 29(4).

137 Id.

before it entered into its sub-lease. If the sub-lessor defaults in its lease payments to the head lessor, the sub-lessee's use of the asset would be subject to the enforcement rights of the head lessor.[138]

2.6.2. *Priority among Buyers and Competing Claimants*

Another potential scenario involving competing claimants can arise when a debtor, after granting a security interest in its assets, sells the same assets to a third-party buyer. In the event that the debtor defaults on his obligations to the secured party, the secured party will want to enforce its security interest against the collateral. However, title to the collateral has meanwhile passed to the buyer – and the buyer will want to maintain title, possession, and control over the assets. This contest is once again subject to the simple "first to file" rule. The buyer's interest in the assets will be subject to any international interest that was registered at the time of the sale.[139] Therefore, if the secured party registered its security interest prior to the sale of the collateral, the secured party will have the right to sell the collateral (or otherwise enforce its security interest), even if the collateral is under the control or in the possession of the buyer.[140] On the other hand, if the security interest was not registered, the buyer would be able to retain title and control over the assets (even if the buyer had knowledge of the security interest when the sale took place).[141] Finally, if the secured party registers its security interest after the sale takes place, the buyer will take free of the security interest and can retain ownership.[142]

Another contest might arise between two buyers of the same asset, such as an on-orbit satellite. If the seller first sells the satellite to Buyer A and later sells the same satellite to Buyer B, the party who registers first will have superior title to the satellite. Of course, in order for the second sale to

---

138 Official Commentary, *supra* note 63, at 68.

139 Space Assets Protocol, Art. XXIII(2). The Convention's provision regarding the priority of buyers (Article 29(3)) is replaced by Articles XXIII(1) and XXIII(2) of the Protocol (as explained in Article IV(1) of the Protocol).

140 Whether the buyer will then be able to pursue a remedy against the debtor is a matter of the local applicable law. In the United States under the Uniform Commercial Code, the buyer may be able to sue the debtor for a breach of the warranty of title since a seller makes a warranty that the goods sold are free of security interests. *See* U.C.C. §2-312.

141 Space Assets Protocol, Art. XXIII(1).

142 Id.

be valid at all, the seller must have still had the power to dispose of the satellite (that is, must have retained an interest in the satellite). However, this is an issue that is left to other applicable law.[143] In another example, a buyer who purchases a satellite after a lease interest has been registered with respect to the satellite must take the satellite subject to the rights of the lessee to use the satellite pursuant to the lease terms.[144]

2.6.3. *Subordination*

The Convention allows for parties holding competing international interests to subordinate one interest to the other by agreement.[145] In other words, parties can enter into a subordination agreement that establishes relative priorities in a way that can differ from the priorities otherwise established under the rules of the Convention. For example, consider the situation in which Bank X who had a second-place priority position (because it was second to file) could, by agreement with Bank Y with a first-priority interest, obtain first-priority status. A subordination of this nature may then be registered in connection with the relevant asset.[146] However, registration of a subordination may only be done with the written consent of the party whose interest has been subordinated (which would require the written consent of Bank Y in the foregoing example).[147] Subordination agreements only affect the priority rights of the parties to the agreement. In other words, a party's interest can only be subordinated if it signs a subordination agreement.

If an international interest is assigned to another party and the assignor had previously signed a subordination agreement, the assignee's priority rights are not affected by the subordination agreement unless the subordination was registered prior to the assignment.[148] This protects a lender from unwittingly buying a loan secured by a security interest that has been

143 See Official Commentary, *supra* note 63, at 113 (stating that "national law applies to questions such as whether an agreement was reached and whether the seller had power to dispose.").

144 Space Assets Protocol, Art. XXIII(2).

145 Cape Town Convention, Art. 29(5). Subrogated interests can also be subordinated by agreement. Id. Art. 38(2).

146 Id. Arts. 16(1(e), 20(2).

147 Id. Art. 20(2).

148 Id. Art. 29(5).

subordinated without the lender's knowledge. The lender need only search the registry to see if any subordinations have been filed. If not, the lender can proceed with confidence that the security interest has priority pursuant to the rules of the Convention.

2.7. *The Effects of Installation, Removal, and Docking of a Space Asset*

It is self-evident that the Convention and Space Assets Protocol do not govern rights in items that are not space objects. For example, a security interest in a personal computer is governed by the applicable domestic law. Moreover, the Protocol does not affect any pre-existing rights in an item that is not a space item when that item is installed on a space item. The effect of such installation is left again to the applicable domestic law.[149] For example, Section 9–335 of the U.C.C. provides that any security interest in an "accession" (that is, "goods that are physically united with other goods in such a manner that the identity of the original goods is not lost") will continue in the accession even after it is united with the other goods.[150] Likewise, the effect of removing a non-space asset from a space asset is also left to applicable domestic law.[151]

The analysis is different when one space object is installed on, removed from, or docked with another space object. In that case, the Protocol does address the effect of such installation, removal, and docking on the existing interest. Under Article III of the Protocol, when a space asset is installed on a separate space asset, any existing interest in the installed asset is not affected by the installation.[152] For example, if during the construction of a communications satellite Bank A has a security interest in a transponder that is subsequently installed in the satellite, Bank A's security interest in the transponder will continue. Similarly, the Protocol also makes clear that merely docking one space asset with another in space will have no effect on any party's interest in the space assets.[153] For example, individual modules in a modular space station might be financed separately by different lenders who have security interests in the modules they financed. When those

---

149 Id. Art. 29(7).

150 U.C.C. §§9-102(a)(1), 9-335(a).

151 Cape Town Convention, Art. 29(7).

152 Space Assets Protocol, Art. III(b).

153 Id. Art. III(a).

modules join together in space, the security interests held by the lenders will not be altered or affected in any way by the docking. Likewise, a party's interest in an item which qualifies as a space asset in its own right, but which has been installed on another space object, will not be affected by the removal of the installed item from the principal space asset.[154]

The definition of "space asset" includes "all installed, incorporated, or attached accessories, parts and equipment," which means that any such item installed on a space asset becomes part of that space asset and is subject to the Protocol.[155] For example, consider a satellite operator that installs a camera on a satellite following the creation and registration of an international interest in the satellite in favor of Bank A. If Bank B held a security interest in the camera under the U.C.C. prior to installation, Bank B's security interest in the camera will continue after installation. However, the camera will also become part of the satellite upon installation – and subject to Bank A's international interest in the satellite. Bank A would have priority over the camera since it is subject to an international interest and, under the priority rules of the Protocol, a registered international interest has priority over an unregistered domestic security interest. In a case where a space object is installed in another space object (as is the case with a transponder installed in a satellite), any pre-existing international interests in the transponder will continue after installation, but the transponder will also become subject to any international interest in the satellite as a whole. The priority of these competing interests will be resolved by the Convention's first-to-file rule.[156]

2.8. *Assignment of Associated Rights and Related International Interests*

The sale of loans by the originating bank to another bank is a common occurrence. The sale allows the bank that issued the loan to recoup the money it disbursed by selling the loan (at a discount) to another bank. This second bank then has the right to receive the payments from the debtor pursuant to the terms of the loan. When a loan is sold, the purchasing bank

---

154 Id. Art. III(b).

155 See Official Commentary, *supra* note 63, at 304–305 (discussing the analogous provision of the Aircraft Protocol).

156 Cape Town Convention, Art. 29(1).

generally buys not only the right to collect payments, but also the security interest that secures the loan. This transfer of the security interest allows the second bank to foreclose on the debtor's collateral in the event that the debtor defaults. This ability to enforce a security interest is a valuable component of the loan package. Without the ability to transfer the security interest, the loan would entail greater risk and the selling bank would likely be unable to sell it (or would be forced to sell at a much reduced price). The ability to sell loans together with the related security interest is an important feature of a modern system of finance because it encourages banks to issue loans since banks know that they will be able to quickly recoup the money disbursed, if the bank so desires, by selling the loan.

The Convention recognizes the need for lenders to sell loans and provides the mechanism to carry out not only the sale of the loan, but also the transfer of the related security interest. The Convention allows these transactions by providing for the "assignment" of "associated rights" together with the related international interest. The term "associated rights" is defined in the Convention as "all rights to payment or other performance by a debtor under an agreement which are secured by or associated with the object."[157] This includes payment obligations under a loan that is secured by a space asset – as well as the payment obligations under a title reservation agreement or a lease. An "assignment" can be an outright sale of the loan (or other obligation), but could also be the granting of a security interest by the originating bank in the loan.[158] For example, Venture Fund might issue a loan to Company and secure the loan on Company's space plane. Venture Fund may then borrow money from Bank and that loan could be secured on Venture Fund's right to collect installments on the loan made to Company. If Venture Fund defaults on its loan, Bank could then enforce its security interest by selling Venture Fund's right to collect installments to the highest bidder (or have Company start making payments under the loan to Bank).

2.8.1. *Formalities of an Assignment of Associated Rights*

An assignment of a loan (or other associated rights) automatically transfers to the assignee the related international interest, that is, the security

[157] Cape Town Convention, Art. 1(c).

[158] Id. Art. 1(b).

interest securing the loan, provided that the formalities under the Convention are followed.[159] Parties can also choose to transfer the associated rights by itself independent of the related international interest, but the Convention only applies to the assignments of an associated interest which also transfer the related international interest.[160] The assignment of a loan without the related security interest will result in the transfer of the loan (which will be governed by other applicable domestic law). Assuming the security interest does not secure another loan, it will be subject to discharge.[161] Similarly, an attempt to transfer an international interest independent of the associated rights will be invalid.[162] Such a transfer is rendered invalid because a security interest serves no purpose unless it is connected with the underlying obligation which it secures. Although an international interest cannot be assigned in isolation, it is not necessary for all of the related associated rights to be assigned in conjunction with the transfer of the international interest.[163] For example, a bank may have issued two loans to a company secured on a single satellite. The bank can transfer one of the loans along with the international interest (the security interest in the satellite) to another bank. In this case, there would be two banks receiving payments on two loans – and both loans would be secured by the same international interest in the satellite. The two banks are free to agree how they will enforce the international interest. Some complications could arise since (as explained below) the two banks will share the same priority position with respect to their security interest and so they will have to come to an agreement as to how they will share the collateral (or subordinate one party to the other).[164] Such an agreement is permitted under the Convention provided that the debtor is not adversely affected without its consent.[165]

In order for an assignment of associated rights (and the related international interest) to be valid under the Convention, the assignment must (1) be in writing and (2) enable the associated rights to be identified by referring to the contract under which the associated rights arose.[166]

---

159 Id. Art. 32(1).

160 Id. Art. 32(3).

161 Regarding discharge of an international interest see *supra* Section 2.5.4.

162 Id. Art. 32(2).

163 Id. Arts. 31(2).

164 Id.

165 Id.

166 Id. Art. 32(1); Official Commentary, *supra* note 63, at 241.

When an associated right is being assigned as security, the agreement must also enable the obligations secured by the assignment to be determined (pursuant to the same standard that applies when creating a security interest under Article 7).[167]

2.8.2. *Effect of an Assignment of Associated Rights*

Following the assignment of associated rights, the debtor is required to make all payments (or perform other obligations owed to assignor) to the assignee. For example, if Bank A assigns Debtor's loan to Bank B, Debtor will at that point make its loan payments to Bank B. However, the debtor is not required to make payments to the assignee until (1) the assignor sends the debtor a written notice of the assignment which describes the nature of the associated rights that have been assigned (for example, a loan obligation) and (2) the debtor agrees in writing to make such payments.[168] This second element effectively gives the debtor the ability to prevent the assignment of associated rights. However, the debtor's agreement can be given at any time and need not specify a particular assignee, so counsel to a creditor should try to include a provision in the original loan agreement that has the debtor agree to make payments to (and otherwise perform for) any assignee upon the assignment of the loan obligation.[169] Once notice has been given to the debtor and the debtor agrees to perform, any payments made by the debtor to the assignee will serve to discharge the underlying obligation.[170] The debtor can agree at any time in writing to waive any and all of its defenses or rights of set-off that it might otherwise raise following the assignment (except that the debtor cannot waive its defense based on the fraudulent acts of the assignee).[171]

Unless the parties agree otherwise, the valid assignment of associate rights automatically transfers all international interests related to the assigned rights.[172] The assignee will also enjoy the same priority held by the assignor with respect to the international interest.[173] For example, if in

[167] Cape Town Convention, Art. 32(1). See *supra* Section 2.4.4 regarding the standard for describing the underlying obligations.
[168] Cape Town Convention, Art. 33(1); Space Assets Protocol, Art. XXIV.
[169] Space Assets Protocol, Art. XXIV.
[170] Cape Town Convention, Art. 33(2).
[171] Id. Art. 31(4).
[172] Id. Art. 31.1(a).
[173] Id. Art. 31.1(b).

the foregoing example Bank A had a security interest with a first-priority position resulting from the early registration of the interest, Bank B will also be able to rely on that registration date to establish its first-priority position with respect to other competing claimants.

2.8.3. *Registering an Assignment of an International Interest*

An assignment of an international interest can be registered in the International Registry.[174] The assignee will want to register the assignment in order to gain priority and protect its interests as described in the following section. As with the registration of international interests, the registration of an assignment can be undertaken by either the assignor or assignee, but the other party must consent in writing (and this consent must be transmitted electronically to the Registrar).[175] The space asset that is the subject of the assigned international interest must be described in the same manner required upon the registration of an international interest.[176] An international interest itself does not need to be registered in order for an assignment of the interest to be registered.

Assignments of international interests can also be registered prospectively.[177] However, once a prospective assignment has been registered, the prospective assignee must discharge the registration if the prospective assignor demands discharge in a writing sent to the address stated in the registration prior to the prospective assignee giving value to the prospective assignor (for example, before paying for the purchased loan).[178]

2.8.4. *Priority Issues Related to an Assignment of an International Interest*

The priority rules applicable to international interests also apply to resolve priority disputes between multiple assignees of the same international interest.[179] A registered assignment has priority over an unregistered assignment as well as having priority over any subsequently registered

174 Id. Art. 16(1).
175 Id. Arts. 18(1) & 20(1).
176 Id. Art. 18(1); Space Assets Protocol Art. XXX. See *supra* Section 2.5.2 for the description requirements.
177 Cape Town Convention, Arts. 1(x), 16(1).
178 Id. Art. 25(2).
179 Id. Art. 35(1).

assignment.[180] However, this priority has important limitations. An assignee will not have priority over another assignee if the contract under which the associated rights arise does not state that the rights are secured by or associated with the asset subject to the international interest.[181] For example, if Bank One has issued to the debtor a number of separate loans, all of which are secured on a satellite owned by the debtor, and Bank One assigns one of the loans to Bank Two and subsequently assigns the same loan to Bank Three, the Convention will not govern the priority between Bank Two and Bank Three if the assigned loan agreement does not state that the loan obligation is secured by the satellite. The rationale for this result is that it would not be fair to subordinate Bank Three to Bank Two (assuming Bank Two registered its assignment) under the rules of the Convention because Bank Three would have had no way of knowing that the Convention governed priority in the absence of any statement regarding the satellite – and, therefore, Bank Three would not have known to search the registry to determine whether another assignee had already registered an assignment.

An assignee will also not be subject to the Convention's priority rules to the extent that the associated rights are not "related" to a space asset.[182] In those situations where the Convention's priority rules do not apply to resolve a contest between multiple assignees, other applicable law will determine priority.[183] Pursuant to Article 36(2), associated rights are "related" to a space asset only to the extent that the associated rights consist of rights to payment or performance related to:[184]

- a sum advanced and utilized for the purchase of the space asset;
- a sum advanced and utilized for the purchase of another space asset in which the assignor held another international interest if the assignor transferred that interest to the assignee and the assignment has been registered;
- the price payable for the space asset;
- the rentals payable in respect of the space asset; or
- other obligations arising from a transaction referred to in any of the preceding subparagraphs.

180 Id. Art. 29(1).
181 Id. Art. 36(1)(a).
182 Id. Art. 36(1)(b).
183 Id. Art. 36(3).
184 Id. Art. 36(2).

The first situation arises when a lender disburses funds that are intended to be used – and are *actually* used – to purchase the space asset. The assignment of this loan will be subject to the Convention's priority rules. The second situation involves "cross-collateralization" where a space asset serves as collateral for funds advanced for the purchase not only of that space asset, but of other space assets that are later purchased with additional advances (provided that a separate international interest is created in these other space assets, this separate international interest has also been assigned to the assignee, and the assignment has been registered). The third situation applies to credit purchases where the debtor has purchased a space asset on credit and has an obligation to pay installments toward the purchase price. The fourth situation concerns the payment of rental for a space asset. Finally, the fifth category entails other obligations that arise in connection with any of the foregoing transactions, such as default interest, break-up penalties, and the costs of enforcing remedies.[185]

In some cases, the assignee of an international interest will compete for priority with the holder of another international interest in the same asset. If the assignee's interest was registered prior to the interest held by the competing claimant, then the assignee will prevail. Typically, the original holder of the assignee's interest, the assignor, will have registered the interest. It is this date of original registration of the international interest that will determine whether the assignee has priority over the holder of the competing international interest. Whoever registered first will have priority. However, in some cases the assignor may not have registered the international interest. This would leave the assignee vulnerable to other registered international interests in the asset. In such a case, an assignee should register the assigned international interest immediately upon assignment (or have the assignor register it prior to assignment) to ensure priority over all subsequently registered interests. As an illustration of this concept, consider a scenario where the owner of a satellite has granted a security interest in the satellite in favor of Bank A to secure a loan issued by the bank. Bank A then assigns the secured loan to Bank B. Assume that the owner then takes another loan related to the same satellite from Bank C and grants the bank a security interest in the satellite. Bank B would be wise to make sure that the assigned international interest is registered before Bank C registers its international interest. Otherwise, Bank C would

185 Official Commentary, *supra* note 63, at 250.

enjoy priority over Bank B. An assignee will also want to ensure that its international interest is registered in order to establish priority over any unregistered interests, such as a competing claimant that has a lien against the debtor's assets.[186]

As mentioned above, assignments of international interests may also be registered prospectively. And, as is true for international interests, when the assignment actually takes place, the date of registration for determining the priority of the assignment will be the date of the prospective registration (provided the prospective registration is still valid when the assignment takes place).[187]

2.9. *Assignment of Debtor's Rights*

The Space Assets Protocol expands its scope of application in a way that is not found in either the Aircraft Protocol or the Rolling Stock Protocol by providing for the assignment of so-called "debtor's rights" by way of a "rights assignment."[188] Debtor's rights are defined as "rights to payment or other performance due or to become due to a debtor by any person with respect to a space asset."[189] In other words, the Protocol provides for the assignment by a debtor of any income stream payable to the debtor that is related to a space asset. For example, the owner of telecommunications satellite who has granted a security interest in the satellite to a bank can also assign to the bank the owner's right to collect lease payments from a television broadcaster. Of course, the definition of "debtor's rights" is worded more broadly so as to include not only payment rights, but the rights to any other obligation owed to the satellite owner in connection with the satellite. The concept of a "rights assignment" is also broadly drawn to include not only the assignment of such debtor's rights for purposes of security, but to also include the assignment of an ownership interest in the debtor's rights.[190]

This expansion of the Protocol to include the assignment of debtor's rights is so significant because it brings within the ambit of the Protocol

[186] Id. at 78–79.
[187] Cape Town Convention, Arts. 19.4 & 19.5.
[188] Space Assets Protocol, Art. II(1).
[189] Id. Art. I(2)(a).
[190] Id. Art. I(2)(h).

classes of property that are not contemplated by the Convention. Instead of being narrowly applicable to only those transactions directly involving space assets (that is, the sale of, lease of, and security interest in a space asset), the Protocol applies to transactions involving accounts receivable, payment instruments, and any number of other forms of performance obligations that could be owed to the owner of a space asset. However, as discussed in the following section, there are some criteria that must be met in order for these transactions to be governed by the Protocol.

2.9.1. *Formalities of a Rights Assignment*

In order for a rights assignment to come within the scope of the Protocol, the debtor's rights related to a particular space asset must be assigned in order to secure, reduce, or discharge an obligation that is already secured on (or otherwise associated with) the same space asset.[191] For example, a debtor's assignment of satellite lease payments to a creditor will be governed by the Protocol if the debtor has already granted a security interest in the same satellite to the creditor. In contrast, if the debtor assigns as security this income stream by itself (without a separate international interest existing in the satellite), the Protocol will not apply. Likewise, if a conditional seller of a satellite also assigned its ownership interest in the lease payments to the buyer of the satellite, the assignment would be subject to the Protocol. However, an isolated assignment of the lease payments would not.

The formal requirements for the documentation of an assignment of debtor's rights are kept to a minimum, as is true for the creation of an international interest.[192] First, there must be a written agreement, although as with an agreement creating an international interest, the concept of a "writing" is broadly defined to include any record (including electronic records) that can be reproduced in a tangible form.[193] The writing must enable (1) the debtor's rights to be identified and (2) the space asset related to the debtor's rights to be identified.[194] This does not mean that the document has to contain a specific description of the debtor's rights or the

191 Id. Art. I(2)(h).

192 Id. Art. IX.

193 Id. Art. IX; Cape Town Convention, Art. 1(nn).

194 Space Assets Protocol, Art. IX(a), (b).

space asset. These requirements will be met as long as the information provided in the document allows a reader (for example, a judge) to determine what rights are being assigned and what space asset is associated with the rights. For example, the rather generic description of "all lease payments received by debtor for the use of all telecommunications satellites owned by debtor" would be sufficient. The Protocol also allows explicitly for the assignment of future debtor's rights (that is, rights that have not yet come into existence), which would enable, for example, a satellite operator to assign payment streams for transponder leases that have not yet been concluded.[195]

An additional requirement for a rights assignment is imposed when debtor's rights are assigned to secure an obligation.[196] When making an assignment by way of security, the contract documenting the assignment must also provide information that enables the secured obligation to be determined – although, as is true when creating an international interest, the writing need not state a precise figure nor a maximum amount.[197] The Protocol also provides for a creditor who has been assigned debtor's rights to reassign those rights (or a portion thereof), referred to as a "rights reassignment."[198] Such reassignment could be carried out, for example, by a lender with a security interest in a debtor's satellite who has also been assigned as security the payment stream arising from the debtor's transponder lease agreements. The lender could sell the secured loan and reassign as security to the buyer of the loan the transponder lease payment streams. Subsequent assignments of the same rights can also take place.[199] Such reassignments are subject to the same provisions of the Protocol that apply to original assignments of debtor's rights.[200]

2.9.2. *Effect of a Rights Assignment*

The extent to which a rights assignment is effective in transferring the rights to the assignee is not determined by the Protocol, but is left to other

195 Id. Art. XI.
196 Id. Art. IX(c).
197 See *supra* Section 2.4.4 regarding this requirement in the context of creating an international interest.
198 Space Assets Protocol, Arts. I(2)(i), XV.
199 Id. Art. I(2)(i).
200 Id. Art. XV(1).

applicable law.[201] Assuming that the rights are transferred, the obligor (for example, a communications company that is making the assigned payments under a transponder lease) is bound by the assignment and is required to make payments to the assignee (or perform any other obligations that are the subject of the rights assignment) provided that the debtor has given the obligor written notice of the rights assignment that identifies the rights that have been assigned.[202] This notice can also be given by the creditor/assignee if the debtor has defaulted on its obligations to the creditor/assignee.[203] If these conditions are met, then any payment made by the obligor will go towards the discharge of the obligor's obligation.[204] Any defenses or set-off rights available to the obligor (who is obligated to make the payments or otherwise perform) against the creditor/assignee are left to other applicable law, although the Protocol will render enforceable a written waiver by the obligor of all or any of the defenses and rights of set-off (with the exception of defenses arising from fraudulent acts).[205]

2.9.3. *Recording a Rights Assignment*

Just as a creditor will want to ensure the priority of its international interest by registering it, a creditor who is the assignee of debtor's rights will want to "record" the assignment in order to secure the priority of its interest in those rights. The assignee must "record" the assignment by adding a record of the assignment to the registration of the related international interest.[206] The assignee can add this record of the assignment at the time of registration or anytime afterwards by amending the registration.[207] A party who acquires debtor's rights by subrogation can also record this acquisition of debtor's rights.[208] This record of the rights assignment (or acquisition by subrogation) need not specifically describe the nature of the debtor's rights.

201 Id. Art. X(1).
202 Id. Art. XIV(1).
203 Id. Art. XIV(2).
204 Id. Art. XIV(3).
205 Id. Art. X(2) & (3).
206 Id. Art. XII(1). The reassignment of debtor's rights by the creditor (the original assignee) can only be recorded on the registration of the assignment of the international interest to that subsequent assignee. Id. Art. XV(2).
207 Id. Art. XII(1).
208 Id.

A mere statement that the debtor has assigned all or some of the debtor's rights (or that the holder of the related international interest has acquired such rights) is sufficient since it places the public on notice that the holder of the international interest has an interest in the debtor's rights – leaving it to the third party to inquire further of the debtor regarding the particular rights that have been assigned or otherwise acquired.[209] This recordation of a rights assignment is otherwise treated in the same manner as the registration of an international interest. Either the debtor or the assignee can record the assignment, although the written consent of the other party is required and must be submitted electronically to the registry, and the recordation can be discharged only with the written consent of the assignee.[210] The recording of a rights assignment is also subject to any regulations issued by the Supervisory Authority.[211] As with international interests, a recording of a rights assignment is complete upon the entry of the information into the registry so as to be searchable.[212] The rules governing the discharge of a registration, as described above in Section 2.5.4, also apply to the discharge of a recording of a rights assignment.[213] In the event that the registration of the related international interest is discharged, the recording of any rights assignment that is part of the registration is also automatically discharged.[214] A search for the recording of a rights assignment can only be carried out by conducting a search for the particular space asset related to the assigned rights. The search certificate provided by the Registrar is required by the Protocol to contain any information that has been recorded regarding the assignment or acquisition of debtor's rights.[215]

In the event that the related international interest is transferred upon the assignment of the associated rights (in other words, when the right to collect the underlying obligation is assigned to another party), any debtor's rights that are recorded as part of the registration of international interest are automatically transferred along with the international interest to the third party transferee.[216] Moreover, the transferee also has the right to be

209 Id.

210 Id. Art. XII(2); Cape Town Convention, Arts. 18(1)(a) & 20(1).

211 Space Assets Protocol, Art. XII(2); Cape Town Convention, Art. 18(1)(a).

212 Space Assets Protocol, Art. XII(2); Cape Town Convention, Art. 19(2).

213 Space Assets Protocol, Art. XII(2); Cape Town Convention, Art. 25.

214 Space Assets Protocol, Art. XII(5).

215 Id. Art. XII(3).

216 Id. Art. XII(4).

shown on the registration as the assignee of the debtor's rights.[217] The practitioner should note, however, that these rights of the transferee only apply when the prior rights assignment has been recorded.

2.9.4. *Priority among Multiple Assignees*

A debtor may decide to assign the same debtor's rights, such as a payment stream for a transponder, to multiple parties. This can come about if a satellite operator receives loans from multiple banks and provides the banks security by granting them a security interest in the same satellite and, in addition, by giving them each a security interest by way of a rights assignment in the payment streams received from transponder leases. If the operator defaults and each bank attempts to enforce its security interest against the lease payments, the question of priority will arise. The Protocol would grant priority to the bank that recorded its rights assignment first.[218] Although it is possible to record a rights assignment as part of the registration of a prospective international interest, the rights assignment will be treated as unrecorded (thus depriving the assignee of priority) until the prospective international interest ripens into an international interest.[219] At that moment, the recorded rights assignment will be treated as recorded and the date of recording (which occurred while the international interest was only prospective) will be the date used to determine priority.[220] However, the date of recordation will only be used to determine priority if the registration of the prospective international interest was still valid when the international interest arises.[221]

2.10. *Sales*

The Convention does not automatically apply to the outright sale of a space asset (as opposed to a conditional sale), but the Convention does provide for the possibility that outright sales may come within its scope if the relevant protocol provides for this.[222] And like the Aircraft Protocol, the

---

217 Id.
218 Id. Art. XIII(1).
219 Id. Art. XIII(2).
220 Id.
221 Id.
222 Cape Town Convention, Art. 41.

drafters of the Space Assets Protocol took advantage of this opportunity by explicitly extending the operation of the Convention to the sale of space assets.[223] Given the differences between an outright sale and the other transactions governed by the Convention, it may seem inappropriate to extend the application of the Convention to outright sales – and in some cases the provisions of the Convention and Protocol are clearly irrelevant to a sales transaction (such as foreclosure remedies) – but other provisions address the concerns of a buyer without the need to alter the language, other than to exchange the term "sale" for "international interest", "buyer" for "debtor", and "seller" for "creditor" (which is precisely what the Protocol does).[224] For example, a buyer of an on-orbit satellite can ensure, by registering its sale first, that it has superior title to the satellite in the event that the prior owner subsequently sells the satellite to another buyer – just as a creditor with a security interest in a satellite can ensure that it has priority over a competing claimant that later acquires a security interest in the same satellite by registering first. The mechanics of registering the sale of a space asset are the same as when registering an international interest. For example, the asset must be described in the same way that the asset would be described when registering an international interest. Prospective registrations are also permitted so that a buyer can ensure its priority before releasing funds.

The Protocol extends the operation of the Convention to the sale of space assets by providing that certain provisions of the Convention apply to outright sales – and that the terms should simply be treated as if they apply to sales transaction (which means that references to an agreement creating international interest should be treated as references to a sales agreement, while references to an international interest, a prospective international interest, debtor and creditor should be treated as references to a sale, a prospective sale, the seller and the buyer).[225] These provisions include the following:

- Article 1 containing definitions
- Articles 3 and 4 regarding the sphere of application
- Article 5 regarding interpretation of the Convention and applicable law

---

[223] Space Assets Protocol, Art. IV.

[224] Id.

[225] Id. Art. IV(1) & (3).

- Chapters IV through VII regarding the registration system
- Article 16(1)(a) providing for a registry for sales transactions
- Article 19(4) allowing for the date of registration of a prospective sale be used to determine priority when the prospective sale ripens into a true sale
- Article 20(1) requiring consent prior to the registration of a contract of sale or a prospective sale
- Article 25(2) regarding the discharge of a prospective sale
- Article 29 regarding priority (with the exception of Article 29(3) which is modified by Article XXIII of the Protocol)
- Article 30 regarding the effects of insolvency
- Chapter X regarding declarations concerning the registration and priority of non-consensual rights and interests
- Chapter XII regarding jurisdiction (with the exception of Article 43)
- Chapter XIII regarding the relationship of the Convention with other conventions
- Chapter XIV containing the final provisions of the Convention (with the exception of Article 60)

The Convention defines a "sale" as a "transfer of ownership of an object pursuant to a contract of sale."[226] "Contract of sale" is, in turn, defined as "a contract for the sale of an object by a seller to a buyer which is not" a security agreement, a title reservation agreement or a leasing agreement.[227] Although the distinction among these different types of agreements may appear to be clear, this is not always the case. For example, as mentioned above, there may be some question about whether a transaction should be treated as a lease or an installment sale with a security interest.[228] There may be similar uncertainty about whether a transaction qualifies as a conditional sale or an outright sale with a security interest. When determining whether a particular agreement should be deemed a security agreement, leasing agreement or title reservation agreement, Article 2 of the Convention directs the parties to applicable law to make the determination.[229] However, Article 2 does not mention how to resolve uncertainty with respect to the identification of a sales agreement. In the event of any

226 Cape Town Convention, Art. 1(gg).
227 Id. Art. 1(g).
228 See *supra* Section 2.1.4.
229 Cape Town Convention, Art.2(4).

uncertainty, the parties can turn to the general provisions of Article 5, which state that where uncertainty arises with respect to a matter governed by the Convention the matter should be settled in conformance with "general principles" or, if such general principles are lacking, in conformance with other applicable law.[230]

In addition to the sale of a space asset, the Protocol also governs the priority of an assignment of any rights to payment (or any other performance owed to the seller) that are related to the sold asset.[231] Therefore, the seller can transfer to the buyer of a satellite not only the satellite itself, but any rights to payment flowing to the seller, such as payments made by the lessee of the satellite's transponders. The effects of such an assignment, the mechanics of recording the assignment, and the priority rules governing such assignments are governed by the same rules that apply to the assignment of debtor's rights, as described above.[232]

2.11. *Remedies*

The ability of a creditor to efficiently exercise a range of remedies in the event of the debtor's default is a critical component of the Cape Town Convention. The purpose of the Convention is to facilitate transactions by reducing the risk of financial loss for the creditor. Such risk is reduced when the creditor is able to recoup losses by having recourse at minimal cost to the asset involved in the transaction. Most importantly, the creditor must be able to exercise remedies without facing the costs (in time and money) of litigation. The Convention provides this efficient exercise of remedies by allowing, under the general rule, recourse to the asset without a court order. However, it should be noted that a state has the option of making a declaration requiring a court order for any of the remedies contained in the Convention or Protocol.[233] Practitioners and parties should be aware of

230 Id. Art. 5(2).

231 Space Assets Protocol, Art. IV(2).

232 Id.; see also *supra* Section 2.9.

233 Cape Town Convention, Art. 54(2). For example, Cuba has made a declaration under Article 54 that "any resource at the disposition of the creditor, according to any of the stipulations of the Agreement, may be exercised solely with the authorization of the Tribunal." See Status of the Convention on International Interests in Mobile Equipment – Signatures, Ratifications, Acceptances, Approvals, Accessions, available at www.unidroit.org/english/implement/i-2001-convention.pdf.

any such declarations limiting the ability of the creditor to exercise remedies in relevant states prior to entering into a transaction under the Convention. In the event that the creditor would like the assistance of the courts in the exercise of a remedy, the creditor is permitted to seek a court order under the Convention.[234] A creditor may want such assistance if the debtor is uncooperative and threatens to create obstacles to the efficient exercise of a remedy.

Different remedies are provided under the Convention for chargees as opposed to lessors and conditional sellers. This is due to the fact that lessors and conditional sellers retain title to the asset involved in the transaction. Therefore, while the core remedy of the lessor and conditional seller is simply to regain possession of the asset, the chargee's remedies when enforcing a security interest provide a mechanism for the chargee to generate money through the disposition (perhaps by the sale) of the asset – although under some circumstances the chargee may itself take title to the asset to satisfy the debtor's obligation.

Although the Convention gives the creditor powerful rights in the event of default, the Convention likewise makes clear that in the absence of default the debtor shall enjoy quiet possession of the asset free from any interference by the creditor.[235]

2.11.1. *The General Requirement of Commercial Reasonability*

When pursuing a remedy under the Convention, the creditor must comply with only a few technical requirements set out in the text of the Convention (such as the notice requirement prior to a sale or lease of the assets, as described below). However, there is also a general requirement, set forth in the Protocol, that all remedies must be carried out in a "commercially reasonable manner":[236]

> Any remedy given by the Convention in relation to a space asset shall be exercised in a commercially reasonable manner. A remedy shall be deemed to be exercised in a commercially reasonable manner where it is exercised in

234 Cape Town Convention, Arts. 8(2), 10(b).

235 Space Assets Protocol, Art. XXV(1).

236 Id. Art. XVII(1). This general requirement of "commercial reasonableness" found in the Protocol explicitly displaces Article 8(3) of the Convention which applies this requirement more narrowly by requiring it only in the context of the remedies available to a chargee under Article 8(1) and the seeking of interim relief under Article 13.

conformity with a provision of the agreement except where such a provision is manifestly unreasonable.

The jurisprudence on the interpretation of similar concepts of commercial reasonableness under applicable domestic law (as exists, for example, under the Uniform Commercial Code) may help give parties and courts some guidance in determining whether a party is acting in a "commercially reasonable manner."[237] Otherwise, the meaning of this standard will be left to the court which will likely look at customary industry practices when making its determination. As reproduced above, the Protocol creates a "safe harbor" for satisfying the requirement of commercial reasonableness by stating that the standard is met if the remedies are carried out in accordance with a provision of the parties' agreement, unless the provision is "manifestly unreasonable."[238] This is a rather broad harbor that will protect the creditor from allegations of unreasonableness provided that the creditor has been successful in describing the manner of enforcement in the security agreement – assuming that the process described does not rise to the level of being "manifestly unreasonable." Such a provision might state, for example, that a debtor who has granted a security interest in a space plane will, upon default, deliver the space plane to the creditor at a specific hangar under the control of the creditor. It is always advisable to add to the agreement language stating explicitly that the parties agree that this agreed upon remedy is commercially reasonable.

2.11.2. *Default*

Remedies can be pursued only upon the debtor's default.[239] However, as is the case in the Uniform Commercial Code and other domestic laws, the

---

237 The requirement that secured parties dispose of collateral in a "commercially reasonable" manner is found in Section 9-610(b) of the Uniform Commercial Code. However, unlike the Cape Town Convention, the Uniform Commercial Code provides additional guidance regarding the meaning of "commercially reasonable" by providing safe harbors that assure the secured party that a disposition will be deemed to have been made in a commercially reasonable manner if "the disposition is made (1) in the usual manner on any recognized market; (2) at the price current in any recognized market at the time of the disposition; or (3) otherwise in conformity with reasonable commercial practices among dealers in the type of property that was the subject of the disposition." U.C.C. §9-627(b).

238 Space Assets Protocol, Art. XVII(1).

239 Cape Town Convention, Arts. 8(1), 9(1) & 10.

term "default" is not strictly defined in the Convention or Protocol. It is instead left to the agreement of the parties to determine what events or circumstances trigger default.[240] Events of default typically include a list of events or situations that may give some indication that the debtor is suffering financial hardship, such as the debtor's default under any of its obligations (even with third parties) or the failure of the debtor to pay its debts as they become due. However, if the parties fail to agree on the definition of default, the Convention provides that remedies may be pursued upon "a default which substantially deprives the creditor of what it is entitled to expect under the agreement."[241] This definition of default excludes trivial, non-material defaults. The practitioner should be sure to include a provision in the underlying agreement specifying the circumstances giving rise to a default in order to avoid this debtor-friendly backup rule.

2.11.3. *Remedies for the Chargee*

Upon default, the chargee has a right to take possession of the collateral, sell or lease the collateral, and collect income from the continued use of the collateral.[242] Although these remedies constitute the very heart of the legal regime created by the Convention, it should be noted that the chargee can only pursue these remedies if the chargor has agreed in advance.[243] This requirement that the chargor agree to the enforcement of a security interest would appear to create a significant obstacle to the efficient enforcement of a security interest – and render the security interest of little, if any, value. Other progressive secured transactions regimes, such as the Uniform Commercial Code, do not require the consent of the debtor. However, this apparent obstacle can be easily removed by requiring the chargor to agree in the security agreement that upon default the chargor will have the right to pursue all remedies provided by the Convention and Protocol. There is nothing in the Convention that prohibits such prior agreement.[244]

---

240 Id. Art. 11(1).

241 Id. Art. 11(2).

242 Id. Art. 8(1).

243 Id.

244 In fact, the language in Article 9 of the Convention indicates that agreement prior to default was contemplated in the context of Article 8 since Article 9 specifies agreement "after default." Id. Art. 9(1).

Although a court order is not required in order to pursue these remedies (assuming no declaration to such effect has been made under Article 54), a chargee can choose to seek a court order authorizing repossession, sale, lease, or collection of income.[245] A chargee may choose to seek a court order if there is some benefit in doing so, such as in order to preempt a chargor who threatens to seek an order blocking repossession or sale, but such an order is not necessary. This avoidance of litigation is one of the most beneficial provisions of the Convention since it eliminates the costs, in time and money, of going to court (costs which would be passed along to the borrower in the form of higher interest rates or the denial of funding).

2.11.3.1. *Repossessing the Collateral*

Upon default, the chargee has a right to "take possession or control" of the collateral subject to a security interest.[246] The chargee does not take title to the collateral, but only has the right to take possession or control. The purpose of this is threefold. First, the chargee will want to protect the collateral from damage, wear from continued use, or the prospect of being sold to a third party by the chargor. Second, having possession enables the chargor to display the asset to potential buyers or lessors when auctioning or otherwise selling or leasing the asset. Third, having possession or control facilitates the transfer of possession or control to a buyer or lessor and also makes it easier for the chargee to operate the satellite (or other asset) in order to collect the revenue from its use. When taking possession is impossible, as in the case of a satellite in orbit, the option of gaining control will provide some of the benefits just mentioned.

The manner in which the chargee gains possession or control of the collateral is not addressed by the Convention except that it must be done in a "commercially reasonable manner." As explained above, the means of taking possession or control will be deemed commercially reasonable if the parties have agreed to the procedure and the terms are not "manifestly unreasonable." The parties might agree, for example, to allow the chargee access to control codes held in escrow upon default for the purpose of gaining control.[247] The method of repossessing collateral, such as automobiles,

245 Id. Art. 8(2).

246 Id. Art. 8(1)(a).

247 Repossessing a satellite in orbit, although possible in low earth orbit, is expensive and technologically challenging. Satellites have been retrieved in the past by the Space Shuttle, such as the Indonesian Palapas B2 satellite, and similar repossessions may take

has been a flashpoint of litigation under domestic laws for many years. The permissibility of certain methods of repossessing satellites and other space assets that are still on Earth may raise similar issues. For example, whether a chargee can tow away a space plane located outside at a spaceport or whether the chargee can open the doors to a hangar and tow a space plane to the chargee's warehouse. Whereas other legal regimes place various types of limits on such behavior, such as the requirement under the Uniform Commercial Code that the secured party not "breach the peace" when repossessing collateral, a chargee operating under the Convention need only ask whether the behavior is "commercially reasonable."

2.11.3.2. *Selling and Leasing the Collateral*

Upon default, the chargee has the right to collect the moneys owed by the chargor by selling or leasing the collateral and applying the proceeds to satisfy the underlying obligation (provided that a declaration under Article 54 of the Convention prohibiting the lease of the collateral has not been made by the state where the collateral is situated in or controlled from).[248] The prerequisites to selling or leasing the asset are few. First, the chargor must agree to the exercise of these remedies.[249] However, this assent can be acquired in the security agreement that the chargor signs at the outset of the transaction. Second, the chargee must give "reasonable" prior written notice of the sale or lease to certain "interested persons."[250] The Protocol provides a safe harbor for the satisfaction of the "reasonableness" requirement by establishing that notice provided fourteen or more calendar days prior to the sale or lease shall be deemed "reasonable."[251] The parties may also agree that a longer period of time is required.[252] The "interested persons" to whom notice must be given include the debtor and any guarantor of the debtor's obligation (a concept that is broadly defined as "any person who, for the purpose of assuring performance of any of the obligations in favour of the creditor, gives or issues a suretyship or demand guarantee or a standby letter of credit or any other form of credit insurance").[253]

---

place in the future. Ribbelink, *supra* note 23, at 43. The repossession of satellites in higher orbits is still a feat of imagination. Id. at 39.

248 Cape Town Convention, Arts. 8(1)(b), 54(1).

249 Id. Art. 8(1).

250 Id. Art. 8(4).

251 Space Assets Protocol Art. XVII(2).

252 Id.

253 Cape Town Convention, Arts. 1(m)(i), (ii), 8(4)(a).

Notice must also be given to any person who (1) has rights in the asset and (2) has "given notice of their rights to the chargee within a reasonable time prior to the sale or lease."[254] In the case of a sale, the buyer will take the asset free of any competing interest in the asset over which the chargee had priority.[255] This means, for example, that no junior secured party will be able to enforce its international interest in the asset following the sale. However, a senior secured party that had priority over the chargee selling the asset will still be able to enforce its international interest against the asset.

As is true for the exercise of any remedy, the chargee must proceed with the sale and lease in a "commercially reasonable" manner.[256] The chargee should take this into consideration when deciding how to sell the collateral (for example, when determining whether to sell the collateral by public auction or a privately arranged sale). In order to make the proper decision, the chargee should look to customary practice in the sale of the type of asset being sold.

2.11.3.3. *Collecting Income Flowing from Collateral*

In addition to repossessing, selling, or leasing the collateral, a chargee also has the right after the debtor defaults to collect "any income or profits arising from the management or use of" the asset.[257] This does not provide the chargee with the right to take control of the asset and operate it for profit. Instead, the chargee only has the right to any income stream that flows from the operation of the asset by the debtor. For a telecommunications satellite, this income would likely take the form of payments on transponder leases. Income related to remote sensing satellites might include payments made by third parties for photographs or other sensed data. This remedy provides a valuable option to a chargee in the event that the sale or lease of the asset is not possible, must be postponed, or is otherwise delayed due to market conditions or other reasons. The Convention does not make clear how the requirement that this remedy be exercised in a "commercially reasonable manner" will affect the pursuit of this remedy. A chargee would be well advised to refer to local applicable law and local practice regarding notice requirements or other protocols that should be

254 Id. Arts. 1(m)(iii), 8(4)(b).

255 Id. Art. 9(5).

256 See *supra* Section 2.11.1.

257 Cape Town Convention, Art. 8(1)(c).

followed when collecting revenue streams in the context of a secured transaction.

2.11.3.4. *Distribution of Proceeds*

Upon the exercise of the above-described remedies, the chargee is required to apply the proceeds to the discharge of the secured obligation.[258] The proceeds can also be applied to cover any reasonable costs arising from the enforcement of the security interest, which might include the costs of warehousing repossessed assets, publicizing and holding an auction, and other incidental costs.[259] Any surplus remaining after the discharge of the underlying obligation and the reimbursement of enforcement costs must be distributed to junior secured parties who have registered their international interest (or have otherwise given notice of their interest to the chargee) in their order of priority.[260] The priority of these claimants may involve the application of applicable domestic law in the event that a priority contest involves two or more unregistered international interests. Any surplus remaining after all claimants have been paid must be returned to the chargor.[261]

2.11.3.5. *Vesting of Interest in Chargee*

An additional remedy that is available to the chargee (holding a security interest) upon the debtor's default is for the chargee to take title to the collateral in satisfaction of all, or part, of the underlying obligation.[262] Article 9 of the Convention provides for the debtor's interest in the collateral to vest in the chargee provided that all "interested persons", as defined in the Convention, agree to the vesting.[263] "Interested persons" include (1) the debtor, (2) any guarantor of the debtor's obligation (as mentioned earlier, this is a concept that is broadly defined as "any person who, for the purpose of assuring performance of any of the obligations in favour of

258 Id. Art. 8(5).

259 Id. Art. 8(6).

260 Id. Article 8(6) contains the proviso "[u]nless otherwise ordered by the court", which appears to invite courts to alter the process of distributing proceeds described in the Article. The role of the court should be limited to interpreting the priority rules under the Convention and applicable domestic law in the event that a party's priority is challenged.

261 Id.

262 Id. Art. 9(1).

263 Id.

the creditor, gives or issues a suretyship or demand guarantee or a standby letter of credit or any other form of credit insurance"), and (3) any person who has rights "in or over" the assets.[264] In order to document this agreement, the chargee should have an agreement drawn up to clarify that the chargor's interest in the collateral shall vest in the chargee in satisfaction of the underlying obligation. The agreement should state precisely what portion of the underlying obligation will be satisfied by the vesting of interest. The risk here lies in identifying all of the parties that have any interest in the collateral. There is no requirement that a party need give notice of its interest – and so this burden remains with the chargee.

The chargee also has the option to go to court and seek an order vesting the chargor's interest in the collateral in the chargee.[265] This has the clear advantage over the method of seeking the agreement of all interested parties since the court order can still be issued over the objection of one or more interested persons. Moreover, the chargee avoids the possibility that a party with an interest in the collateral fails to sign the vesting agreement due to a failure to identify all interested parties. A court can only issue an order to vest "if the amount of the secured obligations to be satisfied by such vesting is commensurate with the value of the object after taking account of any payment to be made by the chargee to any of the interested persons."[266] This requirement ensures that the court does not issue an order that results in a windfall to the chargee (to the detriment of the debtor and other interested parties).

If the collateral is vested in the chargee, the chargee will take the asset free of any competing interest in the asset over which the chargee had priority.[267] This means, for example, that no junior secured party will be able to enforce its international interest in the asset following the sale. However, a senior secured party that had priority over the chargee selling the asset will still be able to enforce its international interest against the asset.

2.11.4. *Redemption*

Redemption is the process by which the chargor, or other party, can keep possession and title to its assets after default (that is, stop the chargee from

264 Id. Art. 1(m).
265 Id. Art. 9(2).
266 Id. Art. 9(3).
267 Id. Art. 9(5).

enforcing its security interest) by paying the chargee the amount of the outstanding obligation. A liberal rule of redemption is found in the Convention. Any "interested party", or the chargor, can redeem the collateral after default by paying the chargee the full amount of the outstanding obligation secured on the collateral.[268] The redeemer may, for example, be a surety that has guaranteed the chargor's obligation. The only limit placed on this rule is that redemption must take place prior to the sale of the asset or the issuance of a court order vesting ownership of the asset in the chargee.[269] The redeemer will also take the collateral subject to any lease that has been granted to a third party by the chargee as a remedy following default (or ordered by a court).[270] The redeemer will not have title to the asset following redemption, but is instead subrogated to the rights of the chargee.[271] In other words, the redeemer can sue for the amount of the chargee's obligation and enforce the chargee's international interest in the asset by sale or by any other remedy available to a chargee. As explained in greater detail below, the redeemer will also have the right to register its acquisition of an international interest.[272]

2.11.5. *Remedies for the Conditional Seller or Lessor*

The remedy for a lessor or for a conditional seller upon default is a simple remedy: terminate the agreement and either (1) retake possession of the asset or (2) take control of the asset.[273] Restoring title to the lessor or conditional seller is not necessary because title was never lost. If necessary, the lessor or conditional seller can choose to seek a court order to allow for the lessor or conditional seller to take possession or control of the asset.[274] After terminating the agreement, the lessor or conditional seller is free to use the asset without restriction (or to sell, lease, or otherwise dispose of the asset) and is free of any further obligation to the lessee or conditional buyer.

---

268 Id. Art. 9(4).

269 Id.

270 Id.

271 Id.

272 See *infra* Section 2.12.3.

273 Id. Art. 10.

274 Id.

2.11.6. *Additional Remedies*

The remedies provided under the Convention are not exclusive remedies. Parties are free to pursue any other remedies available under other applicable law (or agreed upon by the parties) in addition to those contained in the Convention and Protocol, such as remedies available under domestic laws.[275] The only limitation placed on this freedom to pursue additional remedies is that parties cannot agree to remedies that are inconsistent with the mandatory provisions of the Convention set forth in Article 15.[276] The mandatory provisions are discussed below in Section 2.11.16.

2.11.7. *Remedies for the Assignee of Associated Rights as Security*

As mentioned above in Section 2.8, the holder of an international interest can assign the associated rights related to the international interest (such as the right to collect installments on a loan) to a third party to secure an obligation owed by the holder to that third party. In the event of the holder's default on the obligation to the third party, the third party has the right to enforce its security interest in the associated rights (and the attached international interest securing the associated rights) in the same way that a security interest in space assets is enforced.[277] Pursuant to Article 34 of the Convention, the remedies contained in Article 8 (possession, sale, lease, and collection of income from the asset) are available to a party enforcing the security interest, with the understanding that certain aspects of Article 8 will not apply to a security interest in an intangible asset, such as rights to payment which cannot be repossessed by the secured party.[278] The option of having the associated rights vest in the secured party pursuant to Article 9 is also available, as are any additional remedies available under local law pursuant to Article 12.[279] Interim relief pending final determination of the claim under Article 13 is also available to the secured party.[280] Finally, the definition of default under Article 11

[275] Id. Art. 12.
[276] Id.
[277] Id. Art. 34.
[278] Id.
[279] Id.
[280] Id.

extends to this secured transaction as does the rule regarding applicable procedural law under Article 14.[281]

2.11.8. *Remedies for the Assignee of Debtor's Rights as Security*

In the event that a debtor assigns its debtor's rights as security in addition to granting a security interest in the associated space asset, the secured party can utilize the same enforcement measures that are generally available to a secured party under the Convention.[282] Pursuant to Article XVIII of the Protocol, the remedies contained in Articles 8 of the Convention (possession, sale, lease, and collection of income from the asset) are available to a party enforcing the security interest, with the understanding that certain aspects of Article 8 will not apply to a security interest in an intangible asset such as the right to collect payments (as described in the previous section).[283] The option of having the debtor's rights vest in the secured party pursuant to Article 9 is also available, as are any additional remedies available under local law pursuant to Article 12.[284] Interim relief pending final determination of the claim under Article 13 is also available to the secured party.[285] Finally, the definition of default under Article 11 extends to this secured transaction as does the rule regarding applicable procedural law under Article 14.[286] These same provisions apply to the reassignment of debtor's rights by way of security.[287]

2.11.9. *Interim Relief Pending Final Adjudication*

Unless a state has made a declaration opting out of this provision, a creditor (whether a secured party, lessor, or conditional seller) may apply to a court for speedy interim relief if the creditor's right to enforce its international interest is subject to the approval of the court.[288] In order to obtain

281 Id.
282 Space Assets Protocol, Art. XVIII(1).
283 Id. Art. XVIII(1).
284 Id.
285 Id.
286 Id.
287 Id. Art. XVIII(2).
288 Cape Town Convention, Art. 13(1).

interim relief, the creditor has to provide evidence of (1) the debtor's default and (2) the debtor's agreement to the imposition of the relief sought.[289] In the event that a state places limitations on the availability of interim relief, such limitations will be described in a declaration contained in the state's notifications of ratification, acceptance, approval, or accession to the Convention.[290] Whether such a declaration has been made by a state where the asset or debtor is situated should be taken into account by the creditor before entering into a transaction.

A creditor may apply to a domestic court for interim relief of any type that the court sees fit, including (but not limited to) one of the following forms of relief explicitly mentioned in the Convention:[291]

- preservation of the asset and its value;
- possession, control, or custody of the asset;
- immobilisation of the asset;
- lease of the asset; or
- management of the asset.

If the court grants the creditor the right to lease or manage the asset, the Convention contemplates that the creditor would receive any income flowing from such lease or management of the asset.[292] Of course, any order for interim relief may also be subject to limitations. Courts are also permitted to impose any terms necessary to protect the interests of any "interested persons"[293] from any harm that might result from the creditor's (1) failure to perform any of its obligations to the debtor or (2) failure to establish its claim against the debtor.[294] On a related note, the Convention also explicitly allows courts to require that notice be provided

289 Id. Art. 13(1). The debtor's consent to interim relief can be given in advance of litigation, such as in the security agreement or other transactional documents.

290 Id. Art. 55.

291 Id. Art. 13(1). That a court may fashion any type of relief see Article 13(4).

292 Id. Art. 13(1)(d).

293 As mentioned above in Section 2.11.3.5, "interested persons" include (1) the debtor, (2) any guarantor of the debtor's obligation (as mentioned earlier, this is a concept that is broadly defined as "any person who, for the purpose of assuring performance of any of the obligations in favour of the creditor, gives or issues a suretyship or demand guarantee or a standby letter of credit or any other form of credit insurance"), and (3) any person that has rights "in or over" the assets. Id. Art. 1(m).

294 Id. Art. 13(2).

to all "interested persons" prior to issuing an order for interim relief.[295] However, such notice is not mandatory under the Convention.

Although Article 13(4) of the Convention states that the duty to observe "commercially reasonable" practices when enforcing a security interest (as required under Article 8) also applies when a court orders interim relief, this duty may appear to be eliminated since Article XVII(1) of the Protocol nullifies the application of Article 8(3) to space assets.[296] However, at the same time that Article XVII(1) nullifies the application of Article 8(3) it also creates an enhanced duty applicable to space assets that demands adherence to "commercially reasonable" practices when a creditor is pursuing any remedy – which would include an interim remedy ordered by the court under Article 13 of the Convention.[297]

Article XX of the Protocol makes some modifications to Convention's provisions regarding interim relief – but these modifications only apply if a state makes the appropriate declaration to opt-in to Article XX. If a state makes a declaration to apply Article XX regarding relief pending the final determination of a claim, the operation of the Convention regarding such relief will then be modified in accordance with Article XX (although only to the extent set forth in the declaration).[298] A declaration to apply Article XX will have two primary effects. First, a court will be required to issue an order for relief within a certain fixed period of time following the creditor's application to the court, with such time period being stated in the declaration.[299] Second, the court will be explicitly permitted to issue an order requiring the sale of the asset and the application of the sale proceeds to the underlying obligation.[300] However, this remedy is only permitted if the debtor and creditor have entered into a specific agreement allowing for the sale (which agreement, from the creditor's perspective, would be best reached at the outset of the transaction in the original transaction documents).[301] Any interest passing to a buyer in the event of such a sale would

295 Id. Art. 13(3).

296 Space Assets Protocol, Art. XVII(1).

297 Id.

298 Id. Arts. XX(1) & XLI(3).

299 Id. Art. XX(2).

300 Id. Art. XX(3).

301 Id. Art. XX(3). Even without a declaration, the court would be able to issue an order of sale since the Convention makes clear that the court is free to select the form of interim relief it deems suitable. Cape Town Convention, Art. 13(4).

pass free of any interest over which the creditor's international interest has priority (and, conversely, would be subject to any interest that is senior to the creditor's international interest).[302] Finally, the application of Article XX would allow for the creditor, debtor, and any "interested person" to agree in writing to exclude the application of Article 13(2) of the Convention which allows for the court to impose terms on an order for interim relief that protect interested persons.[303]

2.11.10. *Remedies on Insolvency*

The Protocol provides for special remedies upon the occurrence of an "insolvency-related event", but only if the state with "primary insolvency jurisdiction" has made a declaration to provide for such remedies under Article XLI(4) of the Protocol.[304] In its declaration, the state can choose to apply one of the two optional sets of insolvency remedies described below. If one of these options is adopted, the state must adopt the chosen remedies in their entirety (and so cannot choose to adopt a modified form of the remedies).[305] If a declaration is made, the declaring state must specify (1) the types of insolvency proceedings to which the chosen remedies will apply and (2) the length of certain time periods relevant to the application of the remedies.[306] If a debtor is subject to insolvency proceedings in multiple states, all states party to the Convention and Protocol are required to apply the insolvency remedies in accordance with the declaration made by the state of "primary insolvency jurisdiction."[307] Of course, the state of "primary insolvency jurisdiction" can also choose not to make a declaration, in which case other applicable law (domestic or otherwise) will govern insolvency.

Under the Protocol, an "insolvency-related event" is defined as (1) the commencement of insolvency proceedings against the debtor in court or (2) the suspension by the debtor of payments to the creditor (or a statement by the debtor of the intention to suspend payments) when the creditor is prevented by law or state action from commencing insolvency

302 Space Assets Protocol, Art. XX(4).
303 Id. Art. XX(5).
304 Id. Art. XXI(1).
305 Id. Art. XLI(4).
306 Id.
307 Id. Art. XLI(5).

proceedings against the debtor or from exercising remedies under the Convention.[308]

The Contracting State with "primary insolvency jurisdiction" is the state where "the centre of the debtor's main interests is situated."[309] This state will be presumed to be the debtor's statutory seat or, failing the existence of a statutory seat, the state in which the debtor is formed.[310] This presumption, however, is rebuttable.[311]

The first set of remedies that a state may choose to adopt, referred to as "Alternative A" in the Protocol, is the more creditor-friendly of the two options. The core obligation under this option is that the debtor (or insolvency administrator) must give possession or control of the collateral to the holder of an international interest no later than the earlier of (1) the end of the "waiting period" which has been established by the state with primary insolvency jurisdiction in its declaration or (2) the date on which the creditor would have received possession or control under other applicable law if no declaration had been made.[312] Similarly, the debtor or insolvency administrator is required to give the possession or control over any debtor's rights assigned to the creditor within the same time limits.[313] Once the waiting period has elapsed, the creditor may freely pursue any of the remedies under the Convention or Protocol.[314] Until such time as possession or control of the collateral or debtor's rights is given to the creditor, the debtor or insolvency administrator must take necessary actions to preserve the value of the collateral "in accordance with the agreement."[315]

---

308 Id. Art. I(2)(d).

309 Id. Art. I(2)(g).

310 Id. In the case of a business entity, the "statutory seat" is the state in which the registered office of the entity is located as set forth in the entity's foundational documents. For a more detailed discussion of this issue under European law see Soriano, M.V. et al., *The European Insolvency Regulation: Law And Practice* 45 (2004).

311 Space Assets Protocol, Art. I(2)(g).

312 Id. Arts. XXI Alternative A(2), XXI Alternative A(4).

313 Id. Art. XXI Alternative A(3). Of course, if the creditor already has possession or control of the debtor's rights, there will be no need to exercise this remedy. See the footnote to this provision in the penultimate draft of the Protocol in *UNIDROIT Committee of Governmental Experts for the Preparation of a Draft Protocol to the Convention on International Interests in Mobile Equipment on Matters Specific to Space Assets, Fifth Session, Rome, 21/25 February 2011*, Appendix XV, UNIDROIT Doc. C.G.E./Space Pr./5/ Report (2011).

314 Id. Arts. XXI Alternative A(9); XXI Alternative A(13).

315 Id. Art. XXI Alternative A(6)(a).

"The agreement" refers to the security agreement, lease agreement, or conditional sales agreement – and such agreement may not contain any provisions that require maintenance of the collateral.[316] On a related note, the Protocol makes clear that this requirement to preserve value does not preclude the continued use of the space asset provided that such use takes place pursuant to "arrangements designed to preserve the space asset and maintain it and its value."[317] The Protocol also provides that in that interim period, before the time limit for transferring possession or control passes, the debtor or insolvency administrator has the right to seek any interim relief available under applicable law, whether international or domestic.[318]

Of critical importance in the context of insolvency is the provision contained in Article 30 of the Convention which states that an international interest is only effective after insolvency if the interest either (1) has been registered in the International Registry or (2) is otherwise effective in insolvency under other applicable law.[319]

In the course of the insolvency proceeding, the obligations of the debtor may, generally speaking, not be modified without the consent of the creditor – although the insolvency administrator may terminate the debtor's agreement with the creditor if permitted under other applicable law.[320] In the distribution of assets, priority must be given to registered international interests over all other interests, with the exception of those non-consensual rights or interests that have been declared under Article 39(1) of the Convention.[321]

The debtor or insolvency administrator may retain possession and control of the space asset and related debtor's rights beyond the stated time limit in the event that all defaults have been cured (apart from a default resulting from the commencement of insolvency proceedings, which cannot be unwound) and assurance has been given that all future obligations owed to the creditor will be performed.[322] Once transfer of possession and

---

316 Cape Town Convention, Art. 1(a).

317 Space Assets Protocol, Art. XXI Alternative A(7).

318 Id. Art. XXI Alternative A(6)(b).

319 Cape Town Convention, Arts. 30(1) & (2).

320 Space Assets Protocol Arts. XXI Alternative A(10), XXI Alternative A(11).

321 Id. Art. XXI Alternative A(12). See *infra* Section 2.12.4 regarding non-consensual rights or interests.

322 Id. Arts. XXI Alternative A(2); XXI Alternative A(3); XXI Alternative A(8).

control has been averted due to such assurances of future performance, no further grace period will be provided in the event that the future obligations are not met.[323] Any further default will require immediate transfer of possession or control. Possession and control may also be maintained by the debtor or insolvency administrator if transfer is not permitted by domestic export control laws or laws prohibiting the transfer or granting of a license to the debtor or insolvency administrator pursuant to Article XXVI of the Protocol.[324]

Under the second option, Alternative B, a state may choose a significantly less creditor-friendly set of remedies. The core remedy under this alternative only requires the debtor or insolvency administrator, upon the occurrence of an "insolvency-related event", to notify the creditor within the time period established by the state in its declaration that either (1) all defaults, other than the initiation of an insolvency proceeding, will be cured and that all obligations will be performed or (2) the creditor will be permitted to take possession of or control over the space asset in accordance with other applicable law.[325] The difference from Alternative A is stark. Although there is still a time limit by which the debtor or insolvency administrator must act, control over the asset need not be transferred by the expiration of the time limit. Instead, the debtor or insolvency administrator need only give notice by the end of the time period. When possession and control are actually transferred is a matter of other applicable law. Under Alternative B, the creditor is required to provide the court with proof of its claims and its registered international interest and the court is permitted to impose other requirements prior to the transfer of control as may be required under the applicable domestic law.[326] This alternative set of remedies also prohibits the sale of the asset pending a final decision of the court.[327] Alternative B permits a court to transfer possession of, or control over, the asset in the event that the debtor or insolvency administrator fails to give the required notice or fails to transfer promised possession or control.[328] However, the court is not required to do so and is permitted to

323 Id. Art. XXI Alternative A(8).

324 Id. Arts. XXI Alternative A(2); XXI Alternative A(3); XXVI(2). Regarding the limitation on remedies arising from export control and licensing laws see *infra* Section 2.11.14.

325 Id. Art. XXI Alternative B(2).

326 Id. Arts. XXI Alternative B(3); XXI Alternative B(4).

327 Id. Art. XXI Alternative B(6).

328 Id. Art. XXI Alternative B(5).

impose any requirements on the creditor it sees fit prior to transferring possession or control.[329] This stands once again in stark contrast to Alternative A which, upon the expiration of the waiting period, allows the creditor to pursue any remedy without requiring a court order.

The Protocol also requires that assistance be given by the courts of certain states other than the state of primary insolvency jurisdiction in the enforcement of remedies upon insolvency – but only if a state has made a declaration to this effect.[330] If the declaration has been made, the declaring state must "co-operate to the maximum extent possible" in the enforcement of the insolvency remedies, but need only do so in accordance with its own laws.[331] A state need only provide such assistance if the state has one of the following connections to the transaction:[332]

- The space asset is "situated" in the state;
- The space asset can be controlled from the state;
- The debtor is located in the state;
- The space asset is registered in the state;
- The state has issued a license with respect to the space asset; or
- The state has some other close connection to the space asset.

A space asset will be deemed "situated" in a state not only when the asset is physically located in the territory of the state on Earth, but also when the asset is in space if (1) the asset has been registered by the state under the Outer Space Treaty, the Registration Convention, or the 1961 UN Resolution for International Cooperation in the Peaceful Uses of Outer Space (General Assembly Resolution 1721 (XVI) B of 20 December 1961), (2) the state has issued a license for the operation of the space asset, or (3) the space asset is controlled via a mission operation center on the territory of the state.[333] The assistance of other states that have these connections to the space asset will obviously be of great value when a court or insolvency administrator is attempting to transfer control or possession to the creditor.

As a final note, the Convention makes clear that its provisions do not affect any domestic laws regarding (1) the avoidance of a transaction as a preference or transfer in fraud or (2) procedure applicable to the

329 Id. Art. XXI Alternative B(5).
330 Id. Arts. XXII(1); XLI(2)(b).
331 Id. Art. XXII(2).
332 Id. Art. XXII(2).
333 Id. Art I(4).

enforcement of rights against property that is under the control of the insolvency administrator.[334]

2.11.11. *Command Codes and Related Data*

When a creditor enforces an international interest by, for example, selling or leasing the asset, enforcement will be facilitated if the parties place in escrow all command codes and other related data needed to gain control over and operate the space asset. The benefits of this are significant and even necessary. For example, it would be impossible for a chargee to sell an on-orbit satellite if the chargee did not have the control codes needed to operate the satellite. Article XIX of the Protocol allows for the parties to place these data and materials in escrow.[335] However, the Protocol does not require this nor does it give a creditor the right to access this important data in the absence of an agreement. The Protocol simply recognizes the right of parties to agree. This provision creates no additional rights or obligations, but only restates existing law that allows parties to enter into an agreement. The Protocol also makes clear that the right of the parties to enter into such agreements is still subject to any domestic restrictions placed on the transfer of command codes or other such data under export control laws or other regulations.[336]

2.11.12. *Procedural Requirements for the Enforcement of Remedies*

The administrative or judicial procedural requirements for the enforcement of a remedy under the Convention or Protocol will be determined according to "the law of the place where the remedy is to be exercised."[337] In connection with this rule, it should be noted that a state has the option of making a declaration requiring a court order for any of the remedies contained in the Convention or Protocol.[338] Notwithstanding the foregoing, if the state of primary insolvency jurisdiction has adopted Alternative A under the Protocol's article on insolvency remedies, the rule prohibiting the prevention or delay of the exercise of remedies following the expiration

334 Cape Town Convention, Art. 30(3).

335 Space Assets Protocol, Art. XIX.

336 Id.

337 Cape Town Convention, Art. 14. See also Official Commentary, *supra* note 63, at 192.

338 Cape Town Convention, Art. 54(2).

of the applicable waiting period will not allow procedural requirements that obstruct or delay the insolvency remedy.[339]

2.11.13. *Limitations on Remedies: Physically Linked Space Assets*

The modular nature of certain space assets combined with the inability to easily disassemble space assets (particularly when the assets are in orbit) may result in a situation where the enforcement of an international interest in one space asset may have a detrimental effect on the operation of another space asset. For example, a modular space station comprised of separately financed modules (similar to the construction of the International Space Station) may give rise to a scenario where one creditor has a security interest in one module, while another creditor has a security interest in another module. In the event that one of the creditors needs to enforce its security interest by selling Module A to a new owner, the operation of Module B may be harmed if, for example, Module B can no longer independently operate because propulsion and control equipment is located on Module A (perhaps Module B is a laboratory module that was not equipped with independent propulsion and control systems). This could harm the interests of the creditor with a security interest in Module B if the new owner of Module A detached the module and left Module B adrift in space and subject to collisions or uncontrolled reentry.

In order to address this issue, a provision was originally proposed which prohibited the exercise of any remedy against one space asset that would impair ownership rights in another space asset to which it was physically linked. Members of the satellite industry voiced concerns over this proposal since it would have effectively granted a right of quiet enjoyment to owners of, or creditors with security interests in, satellite components. The concern arose from the possibility that a creditor with a security interest in a satellite would be prevented from enforcing that security interest if the remedy would interfere with the operation of any components of the satellite (such as hosted payloads). It was feared that this impediment to pursuing remedies would deter the financing of satellites. The alternative approach recommended by the industry was to leave the issue of enforcing remedies with respect to linked assets to intercreditor agreements, as is

[339] Space Assets Protocol, Art. XXI Alternative A (9). See also Official Commentary, *supra* note 63, at 192.

current practice. The drafters ultimately struck a compromise in Article XVII of the Protocol which allows for the use of intercreditor agreements to limit the application of remedies if the parties involved agreed to do so. However, in the absence of such an agreement, Article XVII prohibits a creditor from enforcing an international interest with respect to a space asset in the event that (1) the asset is "physically linked" to another space asset, (2) the enforcement of the international interest would "impair or interfere with the operation" of the other asset, and (3) another international interest or sale was registered against the other space asset before the registration of the international interest that is being enforced.[340] In other words, although the default rule is that a remedy cannot be pursued if it impairs the operation of another space asset to which it is physically linked, this limitation only applies if the other space asset is subject to a previously registered interest.

When applied to the example described above, the creditor would not be able to sell the propulsion module if the other creditor had registered its international interest in the laboratory module before the international interest in the propulsion module had been registered. The prior registration of the interest in the propulsion module would have put the subsequent creditor on notice of the existing international interest. Therefore, the creditor would know in advance that it could not easily enforce an international interest against the laboratory module should the creditor decide to finance the construction of the module. If the two modules were being financed separately and built simultaneously, the two creditors would be well advised to try to register their own international interest first in order to take advantage of this provision of the Protocol – or enter into an intercreditor agreement to clarify the parties' ability to enforce remedies that might affect the other party.

To add another wrinkle to this scenario, what if the propulsion module had been financed and placed into space before the Cape Town Convention and Protocol entered into force – but soon after the Convention and Protocol enter into force, the laboratory module is constructed and connected to the propulsion module and an international interest in the laboratory module is registered? In this case, the Protocol allows for the creditor to still enjoy the protections of Article XVII provided that the creditor with the interest in the propulsion module registers its international interest

340 Space Assets Protocol, Art. XVII(3).

in the propulsion module within three years of the "Effective Date" of the Convention and Protocol.[341] Provided that a timely registration is made, the creditor with a security interest in the propulsion module will be treated as if it had an international interest registered as of the date that the security interest arose, thus allowing the creditor to enforce this international interest even if it impairs the operation of the laboratory module.

2.11.14. *Limitations on Remedies: Controlled Technology & Licenses*

During the drafting of the Space Assets Protocol, the members of the Working Group were concerned about the transfer of controlled technology that could take place upon the enforcement of remedies under the Convention.[342] The potential military value of space technology has resulted in the close supervision of the transfer of such technology by many states – and particularly by those states that participate in the Wassenaar Arrangement.[343] The laws of a particular state may require a license prior to the export of a space asset or the transfer of a space asset to a foreign national.[344] The drafters ultimately decided to allow states to retain the

341 Id.

342 See, e.g., Lochan, R., "Cape Town Convention & Space Protocol: A Critical Analysis", in Gopalakrishnan, V. and Lochan, R. (eds.), *Conference Proceedings of the ISRO/IISL Space Law Conference 2005: Bringing Space Benefits to the Asian Region*, 6–39, 6–51 (2006).

343 States participating in the Wassenaar Arrangement impose controls on the transfer of military and dual-use to foreign nationals. For example, the transfer of most space technology is controlled in the United States under the International Traffic in Arms Regulations (ITAR) which require the exporting entity to obtain a license from the Department of State prior to the export of such technology. International Traffic in Arms Regulations, 22 C.F.R. §123.8. Regarding the application of the ITAR to space technology see, generally, Sundahl, M.J., "Space Tourism and Export Controls: A Prayer for Relief, 75 Journal of Air Law & Commerce 581 (2010). The 2013 National Defense Authorization Act returned to the executive branch the power to remove satellites from the United States Munitions List (which lists those defense articles that are subject to the ITAR) and subject this technology instead to the less strict Export Administration Regulations (EAR) administered by the Department of Commerce. *National Defense Authorization Act for Fiscal Year 2013*, H.R. 4310 (112th Congress) §1261. It is expected that the President will exercise this authority and transfer commercial satellites to the EAR in the near future.

344 For example, the ITAR require that a license from the Department of State be obtained prior to the mere transfer of control of a satellite to a foreign national. International Traffic in Arms Regulations, 22 C.F.R. §123.8.

control over the transfer of such technology regardless of a creditor's right to effect a transfer under the Convention by allowing a state's courts to refuse to enforce an international interest if such enforcement would conflict with its export controls or national security laws.[345] Even when a court is not involved with the enforcement of a remedy that results in, for example, the sale of a controlled space asset to a foreign purchaser, the Convention should not be interpreted to permit this transfer without the procurement of the required licenses. Such an interpretation would eviscerate export control laws and would allow controlled military technology to fall into dangerous hands.

The Protocol also explicitly preserves a state's right to regulate (1) the "launch or operation of space assets", (2) the provision of services that rely on the use of space assets, such as telecommunications services, (3) the use of orbital positions and frequencies, and (4) the placement of command codes.[346] As a result, if a creditor sells a debtor's communications satellite to a new operator following debtor's default, the debtor's use of that satellite could be contingent on the granting of a license by the state that has the right to regulate the operation of the satellite. Moreover, the Protocol preserves a state's right to place limitations on the transfer of any "licences, approvals, permits or authorisations" that may be held by the debtor.[347]

It should be noted that the Aircraft Protocol does not contain a provision regarding controlled technologies of the type found in the Space Assets Protocol. This may become an issue when handling a transaction involving a vehicle that is only temporarily in space, which transaction would be subject to the Aircraft Protocol. However, this does not mean that export control laws need not be adhered to and licenses for transfers would not be required. The Cape Town Convention was only intended to cover those aspects of a transaction related to commercial law and not to displace other aspects of law, such as export controls.

---

345 Space Assets Protocol, Art. XXVI(3).

346 Id. Art. XXVI(1), (2)(b) & (2)(c).

347 Id. Art. XXVI(2)(a). The term "licence" is defined as "any permit, authorisation, concession or equivalent instrument that is granted or issued by, or pursuant to the authority of, a national or intergovernmental or other international body or authority, when acting in a regulatory capacity, to manufacture, launch, control, use or operate a space asset, or relating to the use of orbital positions or the transmission, emission or reception of electromagnetic signals to and from a space asset." Id. Art. I(2)(e).

2.11.15. *Limitations on Remedies: Public Services*

Another concern unique to the Space Assets Protocol is the deleterious effect that the enforcement of the Convention's remedies might have on public services, such as communications, navigation, air traffic control, tele-medicine, tele-education, natural disaster warning, and disaster management. For example, consider a scenario in which a debtor owned and operated a satellite subject to a security interest that played an integral role in a state's air traffic control system and the debtor defaulted. The creditor might sell the satellite to another operator who could decide to move the satellite to another orbit and thus disrupt the operation of the air traffic control system. This concern was so significant among government representatives on the Space Assets Working Group that a "public services limitation" was demanded in order to protect against the interruption of vital public services that rely on satellite data. In its early incarnation, this limitation was drafted broadly to allow a state to impose any limitations it wished on the exercise of remedies with respect to assets involved in the provision of public services. This created consternation among industry representatives due to the open-ended nature of the limitation. In order to strike a proper balance between the need to protect the public interest while not eviscerating a creditor's remedies, Article XXVII of the final version of the Protocol states that if a public service notice is registered, a creditor cannot enforce any remedy against a space asset if the enforcement of that remedy would render the space asset "unavailable for the provision of the relevant public service" until (1) the creditor registers a notice with the Registry stating that the creditor may take action to enforce remedies if the debtor fails to cure the default before the expiration of a stated cure period and (2) the cure period elapses without default being cured.[348] Article XXVII establishes a number of limits on the application of this stay on remedies that were intended to minimize the economic impact on the creditor.

First, the stay applies only to the enforcement of an international interest in a space asset that provides a public service that is described in a registered "public service notice."[349] This notice must be registered by the "public services provider" that provides the service at issue (pursuant to a

[348] Id. Arts. XXVII(3) & (4).

[349] Space Assets Protocol, Art. XXVII(1).

contract with the debtor or an entity controlled by the debtor).[350] A "public service provider" must be (1) a governmental entity that provides a public service, (2) a private entity situated in the signatory state that has been designated by the state as a public service provider, or (3) an entity that is recognized as a public services provider under the law of the signatory state.[351] The public service notice must describe the nature of the services that have been contracted for between the public services provider and the debtor (in addition to any other requirements imposed by the registration regulations).[352] Registration of the public service notice can take place at any time; however, such registration can only take place if the debtor, public service provider, and the signatory state agree to allow registration of the notice.[353]

Second, the stay on the enforcement of remedies will not apply to international interests that were registered prior to the registration of the public service notice (unless all parties agree that the stay will apply).[354] However, this limitation on the stay only applies if (1) the international interest was created before the debtor entered into the public service agreement, (2) the creditor did not know of the service agreement when registering its international interest, and (3) the relevant public service notice was registered more than six months after the launch of the relevant space asset.[355]

Third, the stay on remedies only prevents the exercise of remedies that impair the provision of the public services. This means that upon default of the debtor a creditor would still be permitted to pursue remedies that did not interfere with the provision of public services. For example, the creditor would be able to collect any income streams flowing from a satellite since this remedy would not interfere with the provision of a public service. And if the public service provider defaults on its obligations to the debtor (such as by failing to make a transponder lease payment), the stay is immediately lifted so that the creditor can pursue the entire range of remedies.[356]

---

350 Id.

351 Id. Art. XXVII(2)(b).

352 Id. Art. XXVII(2)(a).

353 Id. Art. XXVII(1).

354 Id. Art. XXVII(9).

355 Id. Arts. XXVII(9) & (10).

356 Id. Art. XXVII(8).

Fourth, the cure period is limited to a period of time (between three and six months in length) that has been established in a declaration by the state utilizing the public service.[357] This period begins to run upon the registration of the creditor's notice.[358] Following registration of this notice, the creditor must promptly notify both the debtor and the relevant public services provider of (1) the registration of the notice and (2) the date upon which the stay on remedies will expire.[359] If the default is not cured within that applicable period, the creditor is free to pursue the remedies under the Convention and Protocol. In the meantime, the creditor, the debtor and the public services provider are required to cooperate in good faith to find a commercially reasonable way to enable the continuation of the public service (whether such solution involves the use of the space asset or not).[360] In addition, during the stay the government agency of the affected state that issued the license under which the debtor operates the relevant space asset must give the public services provider the opportunity to participate in any proceedings in which the debtor is allowed to participate related to the issuance of a new license permitting another person to operate the space asset.[361] This allows the public service provider to take part in a proceeding that may have a significant effect on the continuation of the public service at issue. Nothing in the Convention precludes the creditor from initiating proceedings, in accordance with the rules of the relevant authorities, to seek another party to replace the debtor as operator of the space asset.[362] During the stay, the creditor may also, if authorized by the relevant authorities, take action to ensure the continued operation of the public service in the event that the debtor is unable to do so.[363]

2.11.16. *Modifying Remedy Provisions by Agreement*

Parties are free to derogate from or modify most aspects of the remedies provided in the Convention and Protocol by agreeing to do so in writing.[364]

---

357 The length of the cure period must by declared by all states upon ratification, acceptance, approval of, or accession to the Protocol. Id. Art. XXVII(4).

358 Id.

359 Id. Art. XXVII(6).

360 Id. Art. XXVII(7)(a).

361 Id. Art. XXVII(7)(b).

362 Id. Art. XXVII(7)(c).

363 Id. Art. XXVII(5).

364 Cape Town Convention, Art. 15; Space Assets Protocol Art. XVI.

However, certain provisions are non-derogable and cannot be modified. Under Article 15 of the Convention, these mandatory provisions include Article 8(3) (the duty to exercise certain remedies in a commercially reasonable manner); Article 8(4) (the duty to give notice to interested persons prior to the sale or lease of an asset); Article 8(5) (the duty to apply all proceeds realized upon the enforcement of a remedy to the underlying obligation); Article 8(6) (the distribution of surplus proceeds); Article 9(3) (requirements of a court vesting ownership of collateral in the secured party); Article 9(4) (redemption rights); Article 13(2) (the imposition of protective terms in a court order for interim relief); and Article 14 (procedural law).[365] Under the Protocol, Article XVII(2) providing a safe harbor for providing reasonable notice prior to sale or lease is non-derogable.[366]

Whether the duty to act in a commercially reasonable manner is mandatory is questionable under the plain language of the Protocol. While the Convention's requirement of the commercially reasonable enforcement of remedies under Article 8(3) is clearly non-derogable under Article 15, Article XVII(1) of the Protocol nullifies the application of Article 8(3) to space assets and substitutes a different requirement regarding commercial reasonability.[367] Since Article XVII(1) is not listed in Article XVI among the mandatory provisions of the Protocol, this duty of commercial reasonability would appear to be subject to derogation or variance.

2.12. *Other Issues*

The following sections address various issues that have not yet been discussed, but which nonetheless play important roles in the operation of the Convention and Protocol. Some of these issues, such as choice of law and the approach to interpreting the instruments, are matters which can broadly affect the application of the Convention and the underlying transaction.

2.12.1. *Proceeds*

One of the risks faced by a secured creditor is that the debtor will sell the collateral to a third-party or that the collateral will be destroyed, thus

365 Cape Town Convention Art. 15.

366 Space Assets Protocol, Art. XVI.

367 Id. Art. XVII(1).

depriving the creditor of its security. This risk is mitigated under the law of secured transactions generally by allowing the security interest to automatically attach to any proceeds realized by the debtor upon the sale of the collateral or to the insurance proceeds received by the debtor. In addition, any priority applicable to the original collateral should be extended to the proceeds so that the creditor will enjoy the same position of priority over competing claimants when it becomes necessary to enforce the security interest against the proceeds. The Uniform Commercial Code provides a good example of a legal system that is particularly liberal with respect to proceeds. Under the Uniform Commercial Code, a security interest automatically attaches to any proceeds received by the debtor – and the nature of proceeds is defined broadly as "whatever is acquired upon the sale, lease, license, exchange, or other disposition of collateral" as well as rights arising from the collateral (such as a right to a dividend payment), any right to pursue a claim arising from harm caused to the collateral, and any insurance proceeds received as a result of damage suffered by the collateral.[368] The Uniform Commercial Code also extends the priority of the original collateral to any proceeds (with certain limitations).[369]

In some respects, the Convention takes a very liberal approach to the treatment of proceeds. Under the Convention, an international interest in an asset automatically attaches to any proceeds that spring from that asset.[370] Moreover, any priority rights related to an asset extend to the proceeds of that asset without any limitation.[371] Therefore, if a creditor has a first-priority security interest in a satellite as a result of being the first to register, the creditor will also have a first-priority security interest in any proceeds of that satellite.

While liberal is the respect described in the foregoing paragraph, the Convention's approach to proceeds is ultimately rather narrow as a result of its constricted view of what qualifies as "proceeds." In contrast to the broad definition of proceeds found in the Uniform Commercial Code, the Convention definition of the term only includes "money or non-money proceeds of an object arising from the total or partial loss or physical destruction of the object or its total or partial confiscation, condemnation

368 U.C.C. §§9-315(a); 9-102(a)(64).

369 Id. §9-315(c).

370 Cape Town Convention, Art. 2(5).

371 Id. Art. 29(6).

or requisition."[372] This definition limits the nature of proceeds to insurance proceeds realized upon the loss or destruction of the space asset, although those proceeds need not take the form of money, but could potentially consist of other forms of payment, such as deposit accounts or financial instruments. The definition of proceeds would also seem to allow for the attachment of the security interest to any compensation received upon a state's expropriation of a space asset, which would qualify as a type of confiscation. This definition of proceeds does not provide for the automatic attachment of a security interest to proceeds received by the debtor upon the sale of the collateral. This reduces the protection available to creditors in the event of such a sale, although the creditor's international interest would still continue in the asset after the sale. A sale does not extinguish the creditor's security interest and, provided the international interest is registered prior to the sale, the creditor would take priority over the rights of the buyer.[373] However, the creditor would not have the option of enforcing its interest against the proceeds of the sale (an option that other legal systems provide, such as the Uniform Commercial Code).

2.12.2. *Salvage Interests*

Insurance is a critical component of any space launch and satellite operation. Insurance is required under national legislation in many cases and provides the financial protection that creditors rely upon when financing a space venture.[374] In order for an insurer to provide insurance at an affordable premium, an insurer will invariably seek salvage rights in the event that the insurer must pay on the insurance policy. Salvage rights can take the form of either "title salvage" or "revenue salvage"[375] An insurer with title salvage rights has the right to take title to a space asset following the payment on the policy, while an insurer with revenue salvage rights has the right to receive revenue generated from the operation of the space asset

---

372 Id. Art. 1(w).

373 Space Assets Protocol, Art. XXIII(2).

374 For example, the United States requires under its launch license regulations that the applicant show financial responsibility calculated under the Maximum Probable Loss analysis. *Financial Responsibility for Licensed and Permitted Activities*, 14 Code of Federal Regulations §440. This showing of financial responsibility is most commonly achieved by the procurement of third-party liability insurance.

375 Meredith, P., "The Space Protocol: Provisions for Insurer Salvage", UNIDROIT Doc. DCME-SP – Doc. 4 (2011).

following a payment on the policy.[376] The custom in the space insurance industry is for insurers to pay the entire agreed-upon value of the satellite in the event of the destruction of the satellite or a "total constructive loss" of the satellite.[377] Under current practice, the threshold for the total constructive loss of a satellite is low and may be realized, for example, upon the failure of only fifty percent of the transponders on a satellite.[378] In light of this, the importance of salvage rights comes into focus. If an insurer pays out an amount equal to the full value of a satellite when it is still operational to a large extent, the insurer's right to take title or receive revenue from the satellite's operation is essential to the ability of the insurer to provide the type of coverage to which the satellite industry is accustomed.

In an early version of the protocol, the inclusion of an insurer's salvage rights within the scope of the protocol caused serious concern in the industry because it gave these rights priority over secured parties in a manner that was not consistent with current practice.[379] This resulted from the protocol's treatment of title acquired through salvage rights as if they were interests acquired by way of a sale. This enabled the interest to be registered and, as a result, gain priority over subsequently registered secured lenders. In addition, upon the discharge of a creditor's debt by the payment of insurance proceeds, the insurer would have been subrogated to any interest held by a creditor in the insured's assets – once again potentially giving the insurer priority over other secured lenders. In a later iteration of the protocol, salvage interests could not be registered, but salvage rights were still subject to effects of the protocol in that they would be deemed "unregistered interests" and, as such, would be subordinate to any registered international interests. In response to this, representatives of the insurance industry voiced the concern that an insurer would risk losing its properly vested salvage rights to a party that had registered an international interest in the same asset.[380] It was argued that this potential disruption would threaten the ability of insurers to continue to offer the

376 Id.

377 Id.

378 Id.

379 Continuing Issues of Concern Regarding the Unidroit Draft Space Protocol to the Cape Town Convention on International Interests in Mobile Equipment dated 16 April 2010, available at www.esoa.net/upload/files/news/unidroit/20100416industry.pdf.

380 See Meredith, *supra* note 375; United States of America, Revenue Salvage – Significant Problems, UNIDROIT Doc. C.G.E./Space Pr./5/W.P. 7 Add. 2 (February 2011); Comments

coverage that is currently relied upon by creditors that finance satellites.[381] In order to avoid the unintended negative consequences of bringing salvage rights into the scope of the protocol, the drafters deleted the controversial provisions and inserted a new provision in Article IV that made clear that the *status quo* with respect to the salvage rights of insurers will not be disturbed by the Convention and that the nature of salvage rights will not be affected by the Convention, but will instead continue to be governed by existing domestic law and contractual arrangements.[382]

2.12.3. *Subrogation*

As discussed above in Section 2.11.4, a party that redeems the debtor's collateral is automatically subrogated to the rights of the secured party. However, the Convention and Protocol also recognize, and allow for, the subrogation to a creditor's associated rights and the related international interest whether such subrogation occurs contractually or automatically through the operation of other law.[383] The subrogee can register the acquired international interest on the International Registry.[384] A party's interest in debtor's rights acquired by way of subrogation may also be recorded when registering the related international interest (or prospective international interest).[385]

2.12.4. *Non-Consensual Rights and Interests*

During the drafting of the Convention, some states insisted that certain rights and interests in collateral held by states (or even by private entities)

---

submitted by Governments, Organisations and Representatives of the International Commercial Space, Financial and Insurance Communities, UNIDROIT 2011, DCME -SP–Doc. 6 (December 2011), available at www.unidroit.org/english/conventions/mobile-equipment/conference2012/dcme-sp-06-e.pdf.

381 Meredith, *supra* note 375.

382 Space Assets Protocol, Art. IV(3). See also Larsen, P.B., "Brief Report on the Final Space Protocol", available at www.iislweb.org/docs/2012_unidroit.pdf (explaining that the Protocol "leaves salvage rights to space assets totally outside the scope of the Space Protocol regime" and that "salvage rights remain subject to national laws.").

383 Cape Town Convention, Art. 38(1). Subrogated rights can also be subordinated through an agreement between the parties that hold competing interests. See *supra* Section 2.6.3.

384 Cape Town Convention, Arts. 16(1)(c), 20(4).

385 Space Assets Protocol, Arts. XII(1), XXXII(1).

should not be subject to the general priority rules under the Convention, but should instead be granted special priority rights. For example, a given state may not want a tax lien that has attached to an asset to be subordinated to a registered international interest through the operation of the Convention. As a result, the Convention includes Articles 39 and 40 which provide that special priority be accorded certain rights and interests that are identified by a state.[386] Under Article 39, a state may make a declaration that certain rights or interests (termed a "non-consensual right or interest") will have priority over any registered international interest.[387] This priority will be granted even in the context of insolvency.[388] However, a state may only make this declaration if the particular right or interest already enjoys priority under its applicable local law over interests that are equivalent to a registered international interest.[389] For example, if a state's domestic law grants seniority to tax liens over any security interest in the same asset, that state would be able to make a declaration under Article 39 to grant tax liens seniority over a registered international interest.[390] The types of non-consensual rights or interests contained in a state's declaration may be either specific or general in nature, that is, they might include a particular tax lien attached to a particular satellite or may include all tax liens that currently exist or will exist in the future.[391]

---

386 Cape Town Convention, Arts. 39 & 40.

387 Id. Arts. 1(s) & 39(1)(a). For example, upon acceding to the Convention, India made a declaration granting priority under Article 39 for liens to cover claims for airline employee wages arising after default, tax liens for arising after default, and liens in favor of repairmen who have completed repairs on aircraft. See Status of the Convention on International Interests in Mobile Equipment – Signatures, Ratifications, Acceptances, Approvals, Accessions, available at www.unidroit.org/english/implement/i-2001-convention.pdf.

388 Cape Town Convention, Art. 39(1)(a); Space Assets Protocol, Art. XXI Alternative A(12).

389 Cape Town Convention, Art. 39(1)(a). This requirement that priority for the right or interest must already exist under domestic law does not apply in every case. When a state or private entity has a right to arrest or detain an asset as a result of an obligation owed for the provision of public services, a state may make a declaration that such right to arrest or detain be given automatic priority over a registered international interest even in the absence of such priority under domestic law. Id. Art. 39(1)(b); see also Official Commentary, *supra* note 63, at 258.

390 Whether this standard of the treatment of equivalent interests under local law has been met may in many cases be debatable.

391 Cape Town Convention, Arts. 39(1)(a), 39(2).

Generally speaking, if a state makes a declaration listing non-consensual rights or interests that will take priority over a registered international interest, priority will only be granted over those international interests that are registered after the declaration is made.[392] This protects the holder of the international interest who registered prior to the declaration and acted (perhaps by extending a loan) in reliance on the fact that it had first-place priority. However, this rule does not apply if the declaration is made at the time of the state's ratification, acceptance, approval of, or accession to the Protocol.[393] A declaration made at such time will allow for priority to be given to the listed non-consensual rights and interests over all international interests, even if the interests were registered prior to the declaration.[394]

In addition to these non-consensual rights and interests that are granted automatic priority, Article 40 of the Convention also allows states to make a declaration listing categories of non-consensual rights and interests that can be registered just like an international interest and, as a registered interest, take advantage of the priority rules as if it were an international interest.[395] When the underlying obligation related to a registered non-consensual right or interest has been discharged, the holder of the right or interest must promptly terminate the registration upon the written demand of the debtor.[396]

The Registry is required to maintain a list of any declarations made regarding non-consensual rights or interests which must be made available to the public and be searchable by country name so that parties intending to enter into a transaction will be able to determine if their international interest is at risk of being superseded by a non-consensual right or interest included in a declaration under Articles 39 or 40.[397]

---

392 Id. Art. 39(3).

393 Id. Art. 39(4).

394 Id.

395 Id. Arts. 16(1)(a), 20, 40. For example, upon acceding to the Convention, Norway made a declaration granting priority to the following registrable rights and interests under Article 40: (1) rights of a person obtaining a court order permitting attachment of an aircraft object in partial or full satisfaction of a legal judgment and (2) liens or other rights of a state entity relating to taxes or other unpaid charges. See Status of the Convention on International Interests in Mobile Equipment – Signatures, Ratifications, Acceptances, Approvals, Accessions, available at www.unidroit.org/english/implement/i-2001-convention.pdf.

396 Cape Town Convention, Art. 25.

397 Id. Art. 23.

2.12.5. *Jurisdiction and Choice of Forum*

The Convention's provisions on jurisdiction reflect a high premium placed on party autonomy. The general rule under Article 42 of the Convention is that parties to a transaction covered by the Convention can select the jurisdiction where any disputes arising under the Convention can be adjudicated.[398] Any such choice of forum agreement between the parties will confer exclusive jurisdiction on the named forum – meaning that should a party attempt to file a suit in a court in another state that is a party to the Convention, the court must dismiss the case for lack of jurisdiction.[399] Of course, the obligation to dismiss such a case would only apply if the state were a party to the Convention. The chosen forum need not have any connection with either of the parties or to the transaction.[400] The only requirements for creating an enforceable forum selection clause are (1) that the selected forum be a party to the Convention and (2) that the agreement be concluded in accordance with the formal requirements of a binding forum selection clause in effect in the chosen jurisdiction.[401] Nothing in the Convention prevents parties from agreeing to arbitrate their disputes, which agreement would be enforceable under the United Nations Convention on the Recognition and Enforcement of Foreign Arbitral Awards.[402]

The exceptions to the general rule regarding the autonomy of parties to choose exclusive jurisdiction arise in the context of (1) interim relief, (2) insolvency, and (3) orders against the Registrar.[403] First, when a creditor's need for swift *in rem* interim relief arises (that is, the need to preserve the value of the asset by taking possession of the asset, immobilizing it, or

398 Id. Art. 42(1). Practitioners should note that a state can make a declaration upon ratification of the Convention to select what particular courts within the state will have jurisdiction over disputes arising under the Convention. Id. Art. 53.

399 Id. Art. 43(1).

400 Id.

401 Id. Art. 43(2). Pursuant to the plain language of Article 43(2), the forum selection clause need not necessarily even be in writing provided that the provision is otherwise enforceable under the law of the selected jurisdiction.

402 Under the New York Convention, if parties have a written agreement to submit a dispute to arbitration, the court of a signatory state must refer the parties to arbitration if a suit is filed. United Nations Convention on the Recognition and Enforcement of Foreign Arbitral Awards, Art. II, June 10, 1958, 21 U.S.T. 2517, T.I.A.S. No. 6997, 330 U.N.T.S. 38.

403 Cape Town Convention, Arts 43, 44 & 45.

taking some other measure – such as parking a satellite in a particular orbital slot – while the creditor is selling the object or pursuing other remedies), the creditor can seek such relief not only in the forum chosen by the parties in their underlying agreements, but also in the courts of the state where the space asset is "situated" (provided that such state is a party to the Convention).[404] In this context, an asset is "situated" in a state if (1) the space asset has been registered by the state under the Outer Space Treaty, the Registration Convention, or the 1961 UN Resolution for International Cooperation in the Peaceful Uses of Outer Space (General Assembly Resolution 1721 (XVI) B of 20 December 1961), (2) the state has issued a license for the operation of the space asset, or (3) the space asset is controlled via a mission operation center on the territory of the state.[405] This broad definition of being "situated" in a state is necessary so that the state that has jurisdiction and actual control over the asset can issue orders to ensure that the value of the asset is protected. Similarly, if a creditor is seeking *in personam* interim relief (such as an order for payments due under a conditional sales agreement), the creditor can seek such relief not only in the jurisdiction agreed upon by the parties, but also in the state where the debtor is situated.[406] Parties should be aware that a state has the right to declare that it will not adhere to jurisdictional rules regarding interim relief.[407] Whether such a declaration has been made by a state where the asset or debtor is situated should be taken into account before entering into a transaction. Second, in the context of insolvency, jurisdiction is granted to the "primary insolvency jurisdiction" given the need for the court overseeing insolvency proceedings to have jurisdiction over the debtor and the debtor's assets as described above in Section 2.11.10.[408] Finally, in the event that a party brings a claim against the Registrar, the

[404] Id. Art. 43(1).

[405] Space Assets Protocol, Art I(4).

[406] Cape Town Convention, Art. 43(2). See also Official Commentary, *supra* note 63, at 264–265.

[407] Cape Town Convention, Art. 55. For example, New Zealand has made a declaration, in accordance with Article 55, that it will not apply the jurisdictional rules regarding interim relief under Article 43. See Status of the Convention on International Interests in Mobile Equipment – Signatures, Ratifications, Acceptances, Approvals, Accessions, available at www.unidroit.org/english/implement/i-2001-convention.pdf.

[408] Cape Town Convention, Art. 45. See *supra* Section 2.11.10 regarding rules of jurisdiction in the context of insolvency.

courts of the state where the Registrar is headquartered will have exclusive jurisdiction to hear such claims.[409] The state where the Registrar is headquartered will also have jurisdiction, and in some cases exclusive jurisdiction, to issue orders for the discharge or amendment of a registration in the event that a party does not respond to a demand to amend or discharge the registration.[410]

When applying the Convention's rules concerning jurisdiction, consideration should also be given to the law regarding jurisdiction over space objects under the existing law of outer space as discussed below in Section 3.6.

2.12.6. *Waiver of Sovereign Immunity*

In the event that one of the parties to a transaction is a state (or a state's subdivision, agency, or state-owned enterprise), that party may well raise the defense of sovereign immunity if faced with a lawsuit in connection with the Convention. If successful, this defense will result in the dismissal of the suit. Although exceptions and limitations to sovereign immunity are found under the various domestic laws, such as the U.S. Foreign Sovereignty Immunity Act, the most effective means by which to eliminate the potential shield of sovereign immunity is to have the party explicitly waive immunity in the transactional documents. The Protocol provides that such a waiver will be effective provided that (1) it is in writing and (2) it contains a description of the relevant space asset that satisfies the requirements applicable to a description of the asset for purposes of creating an international interest.[411]

2.12.7. *Choice of Law*

The Convention and Protocol do not address every issue that may arise in the context of a transaction involving a space asset. For example, the issue of the validity of a contract may arise or a claim may be raised concerning an allegation of fraud. Since the Convention and Protocol do not address such matters, a court would have to look to other applicable law. Although the Convention may not address the substance of a particular legal issue,

---

409 Cape Town Convention, Art. 44(1).

410 Id. Art. 44(2) & (3).

411 Space Assets Protocol, Art. XXXIII. Regarding the requirements for sufficiently describing the space asset see *supra* Section 2.4.3.

the Convention does contain provisions concerning the choice of law which will help guide a court in the selection and application of the appropriate law.

With respect to those issues that are not addressed by the Convention, the Convention directs courts to look first to "general principles on which [the Convention] is based."[412] When it is not possible for a court to identify such "general principles", the Convention requires that the issue at hand be resolved under applicable domestic law under the conflict of laws rules of the forum.[413] In the event that a state is made up of territorial units, such as the individual states of the United States of America, the law of the particular territorial unit applies as determined by either (1) the law of the nation-state or (2) in the absence of such a nation-state law, the law of the territorial unit with the closest connection to the transaction.[414] Notwithstanding these provisions, the Protocol grants the parties to a transaction the right to choose the law that will govern the contracts underlying their transaction.[415] This choice of law will not affect the rights and obligations springing from the Convention and the Protocol, but will govern contractual issues that are not addressed by the Convention or Protocol, such as the validity of a contract or remedies for breach of contract in addition to those provided in the Convention and Protocol. Parties may choose a state's national law to govern their contract or the law of a territorial subdivision of a state.[416] A state party to the Convention can declare that it will not recognize parties' right to choose the governing law.[417] Parties should be aware of this possibility when entering into a transaction and check the declarations made by those states that related to a transaction. If a claim comes before a court in a state that has made a declaration to not enforce choice of law clauses, the parties may be subject to a law that is unfamiliar, at best, and inimical, at worst.

2.12.8. *Interpreting the Convention and Protocol*

The Convention and Protocol, being treaties, must be interpreted in accordance with the rules of interpretation contained in the Vienna Convention

412 Cape Town Convention, Art. 5(2).
413 Id. Arts. 5(2) & 5(3).
414 Id. Art. 5(4).
415 Space Assets Protocol, Art. VIII(2).
416 Id.
417 Id. Art. XLI(2)(a).

on the Law of Treaties.[418] The primary rule under the Vienna Convention is to give the terms of a treaty their "ordinary meaning in their context and in the light of [the treaty's] object and purpose."[419] The "context" of a term consists of the text and preamble of the treaty, while a treaty's "object and purpose" are to be determined from the text of the treaty (and not from external sources of information).[420] As reflected in these rules, the Vienna Convention takes a text-centered approach to interpreting treaties that generally requires strict adherence to the text.[421] In the event that the ordinary meaning of a term is ambiguous (or needs to be confirmed) "supplementary means of interpretation" may be applied to provide clarification.[422] These supplementary considerations include the *travaux préparatoires* of the treaty, such as the Unidroit Working Group documents, as well as the circumstances of the treaty's conclusion.[423]

In addition to these fundamental rules of treaty interpretation, the Cape Town Convention provides its own rules of interpretation which must be considered alongside the Vienna Convention rules.[424] Specifically, Article 5 of the Convention requires that when interpreting the Convention "regard is to be had to its purposes as set forth in the preamble, to its international character and to the need to promote uniformity and

---

418 Vienna Convention on the Law of Treaties, Arts. 31–33, May 23, 1969, 1155 U.N.T.S. 331 [hereinafter Vienna Convention]. Even with respect to those states that have not ratified the Vienna Convention, such as the United States, the interpretational rules of the convention are binding as an expression of customary international law. Gardiner, R.K., *Treaty Interpretation* 14–16, 69 (2008); see also *Legal Consequences of the Construction of a Wall in the Occupied Palestinian Territory* (Advisory Opinion) [2004] ICJ Reports 38, para 94.

419 Vienna Convention, *supra* note 418, Art. 31(1).

420 Id. Art. 31(2). See also International Law Commission, *Draft Articles on the Law of Treaties with Commentaries* 221 (1966); Gardiner at 178–89, 192, 343–45.

421 Gardiner, *supra* note 418, at 144–45; see also Berglin, R.H., *Treaty Interpretation and the Impact of Contractual Choice of Forum Clauses on the Jurisdiction of International Tribunals: the Iranian Forum Clause Decisions of the Iran-United States Claims Tribunal,* 21 Texas Int'l Law Journal 39, at 44 (1986).

422 Vienna Convention, *supra* note 418, Art. 32.

423 Id. Recourse to supplementary considerations is also permitted when the ordinary meaning of a term results in a meaning that is "manifestly absurd or unreasonable." Id. The Vienna Convention also requires that any subsequent state practice that sheds light on the proper application of the treaty be taken into account when determining the ordinary meaning of a term. Id. Art. 31(3).

424 Cape Town Convention, Art. 5(1) & (2).

predictability in its application."[425] To the extent that these rules alter or expand the rules of the Vienna Convention, such an alteration or expansion would supersede the Vienna Convention rules under the doctrines of *lex posterior derogate legi priori* and *lex specialis derogat legi generali.*

2.12.9. *The Convention's Relationship to its Protocols and to Other Treaties*

The Convention and the protocols to the Convention govern areas of law that are addressed by other treaties. In these instances, the question arises as to which treaty should take precedence when deciding issues of law. Although general principles of international law can be applied to resolve this issue, the Convention and its protocols explicitly address their relationship to certain treaties. According to the principle of *lex posterior*, or "later in time" principle, the Convention and its protocols would trump any earlier treaties as a more recent expression of international law. The Convention and its protocols should also supersede any other international law pursuant to the principle of *lex specialis*, which gives supremacy to a law that addresses a specific issue over another law which may handle the issue in a general manner. The following provisions address the Convention's relationship with other certain other treaties. In the first two cases, the supremacy of the Convention is clarified. However, the third provision subordinates the Convention to existing space law treaties.

- Article 45 *bis* provides that the Convention "prevails" over the *United Nations Convention on the Assignment of Receivables in International Trade.*[426]
- Article XXXIV of the Space Assets Protocol provides that the Convention "supersedes" the *Unidroit Convention on International Financial Leasing.*[427]
- Article XXXV of the Protocol states that the Convention "does not affect State Party rights and obligations under the existing United Nations outer space treaties or instruments of the International Telecommunication Union."[428]

---

425 Id. Art. 5(1).

426 Id. Art. 45 *bis*.

427 Id. Art. 46; Protocol Art. XXXIV.

428 Id. Art. XXXV. In connection with Article XXXV, the fourth recital of the Protocol states that the parties are "mindful of the established principles of space law, including those contained in the international space treaties of the United Nations and the

As stated above, in the first two provisions supremacy is accorded to the Convention and the Protocol, while under Article XXXV of the Protocol the Convention appears to have supremacy over existing space treaties. Unfortunately, the variation of language used in each of the above provisions opens the possibility of confusion about whether the different language of these provisions carries different meanings. Article 45 *bis* states that the Convention "prevails", Article XXXIV of the Protocol uses the term "supercedes", and Article XXXV of the Protocol uses the phrase "does not affect State Party rights and obligations." Is there a difference between these terms – or do they each simply mean that the Convention and Protocol trump the other treaties? This can be left to courts to determine if there is any elasticity in this language. The presumption would likely be that these terms mean that the existing space law treaties trump the Convention and Protocol in the event of any conflict. Further discussion of this issue with respect to the space treaties can be found below.[429]

In the event that there is any inconsistency between the Convention and the Protocol, the Convention explicitly states that the Protocol trumps the Convention.[430] This allows the Protocol to more easily modify the Convention in order to tailor the Convention's rules to the particular needs of the space industry (although the primacy of the Protocol could also be argued on the basis of *lex posterior* – since the Protocol will enter into force after the Convention – or *lex specialis* – since the Protocol addresses a more specific subject matter than the Convention, namely, the financing of space assets).

2.12.10. *Declarations*

States have the option to modify their obligations under the Convention and Protocol at the time of ratification, acceptance, approval of, or accession to this Protocol (and, in some cases, at any time) by submitting a declaration stating such modification to the Depositary. A state may also make a declaration after ratifying the Convention by submitting the declaration

---

instruments of the International Telecommunication Union." Id. Fourth Recital. Although recitals are not binding, they do play a role in the interpretation of a treaty as described *supra* in Section 2.12.8.

429 Regarding the relationship of the Convention and Protocol to existing space treaties see *infra* Section 3.2.

430 Cape Town Convention, Art. 6(2).

to the Depositary.[431] In some cases, the state is not required to make a declaration at all (and thereby be subject to a default rule), while in other cases the state must make a declaration regarding the issue, but has discretion regarding the terms of the declaration (such as the declaration under Article XXVII(4) regarding the length of the stay on remedies again assets used in the provision of public services). The issues addressed by these various possible declarations are described briefly in the following list:

- The priority of non-consensual rights or interests (Art. 39)[432]
- The right to arrest or detain a space asset to repay amounts owed to a state agency (Art. 39)[433]
- The ability to register certain non-consensual rights or interests (Art. 40)[434]
- The application of the Convention to internal transactions (Art. 50)[435]
- The application of the Convention and Protocol to territorial units (Art. 52, Art. XXXIX)
- The identification of courts with jurisdiction over issues arising under the Convention (Art.53)[436]
- The granting of a lease of a space asset while it is located in the state (Art. 54(1))[437]
- The requirement that the leave of court is required to exercise a remedy (Art. 54(2))[438]
- The granting of relief pending final determination of a claim (Art. 55)[439]

---

431 Id. Art. 57(1); Space Assets Protocol, Art. XLIV(1). If a declaration is made after ratification, the declaration will take effect on the first day of the month following the expiration of six months after the declaration is received by the Depositary (or at a later date if specified in the declaration). Cape Town Convention, Art. 57(2); Space Assets Protocol Art. XLIV(2). Note, however, that the declaration will not affect any rights and interests that arose prior to the effective date of the declaration. Cape Town Convention, Art. 57(3); Space Assets Protocol, Art. XLIV(3).

432 See *supra* Section 2.12.4.

433 Id.

434 Id.

435 See *supra* Section 2.3.4.

436 See *supra* Section 2.12.5.

437 See *supra* Section 2.11.3.2.

438 See *supra* Section 2.11.

439 See *supra* Section 2.11.9.

- The time period for the stay on the execution of remedies with respect to space assets integral to public services (Art. XLI(1))[440]
- The application of Article VIII regarding choice of law (Art. XLI(2)(a))[441]
- The application of Article XXII of the Protocol regarding the provision of assistance during insolvency (Art. XLI(2)(b))[442]
- The application of Article XX of the Protocol regarding certain aspects of relief pending final determination of a claim (Art. XLI(3))[443]
- The number of days that constitute "speedy" relief in the event that a declaration has been made to apply Article XX of the Protocol[444]
- The application of Alternative A or Alternative B regarding insolvency remedies under Article XXI of the Protocol (Art. XLI(4))[445]

No other reservations limiting the application of the Convention can be made by a state except for these declarations that are explicitly permitted by the Convention and Protocol.[446] This preserves the uniformity of the application of the Convention among the different states, which is a critical feature of any efficient regime of commercial law. In the event that a state has made a declaration and decides to withdraw it, the state may do so, although the withdrawal will not take effect until the first day of the month following the expiration of six months after the notice of withdrawal is received by the Depositary and the withdrawal will not affect any rights and interests that arose prior to the effective date of the withdrawal.[447]

440 See *supra* Section 2.11.15.
441 See *supra* Section 2.12.7.
442 See *supra* Section 2.11.10.
443 See *supra* Section 2.11.9.
444 Id.
445 See *supra* Section 2.11.15.
446 Cape Town Convention, Art. 56(1); Space Assets Protocol, Art. XLIII(1).
447 Cape Town Convention, Arts. 58(1) & (2); Space Assets Protocol, Art. XLV(1).

Chapter Three

The Relation of the Convention to the Law of Outer Space

This final chapter explores the broader field of the international law of outer space with respect to potential intersections or conflicts with the Cape Town Convention and the Space Assets Protocol. The Convention and Protocol are the first new treaties in the field of space law since the Moon Agreement was concluded in 1979 and so there is a great deal of interest (and some concern) on the part of governments and academics regarding how these new instruments will affect the long-standing international law of outer space.[1] However, interest in the intersections between the Convention and existing space law should not be limited to governments and academics. The practicing lawyer must also be cognizant of how the existing international law of outer space may affect the operation of the Convention and Protocol. Practitioners must be aware that the Convention will not be operating in a vacuum, but will be working in a complex environment of multiple space law treaties. The need for this vigilance is heightened by the fact that the Convention and Protocol are, by their own terms, subordinated to the terms of the existing space treaties.[2] In light of this, the practitioner involved in a transaction governed by the Cape Town Convention must understand not only the Convention and Protocol, but also the broader body of space law.

Some of the issues discussed in this chapter are not entirely new. For example, the fact that liability for damage caused by a satellite remains with the launching state even if the satellite has been sold to a company incorporated in another state has been a controversial issue ever since satellites were first bought and sold among private parties. However, the Cape

---

[1] Regarding the Moon Agreement and the other space treaties see *infra* Section 3.1.

[2] See *supra* Section 2.12.9.

Town Convention places greater stress on these areas of space law that are not well suited to the private space industry because the Convention's *raison d'être* is the facilitation of the commercial transactions that create these points of stress. In other cases, the Cape Town Convention has created new tensions in the fabric of existing space law. In particular, certain provisions in the Convention regarding jurisdiction have the potential to create new conflicts with the jurisdictional rules under the Outer Space Treaty and the Registration Convention. In other ways, however, the earlier space treaties may actually assist creditors who need to enforce their rights under the Cape Town Convention. This scenario arises in connection with the Rescue and Return Agreement, which could help financiers recover, for example, a space plane that has landed under distress in a state that is not a party to the Cape Town Convention.

This chapter not only explores these intersections of the Convention and Protocol with existing space law, but also describes in some detail the nature of the relevant aspects of space law. Particular attention is given to highlighting the differences of opinion regarding certain issues in space law where such differences arise. Many issues in space law continue to be subject to debate – and the contours of these debates are described herein with references to the relevant authorities. The resolution of these debates is not the purpose of this book. Instead, the aim here is to explain to the practitioner the different perspectives on the relevant issues so that the practitioner can pursue the line of thought that is appropriate for the case at hand.

3.1. *An Overview of the Law of Outer Space*

The law of outer space takes various forms at multiple levels: international, regional, and domestic.[3] The international law of outer space consists of multilateral and bilateral treaties, and customary international law. Regional laws affecting outer space activities (which can also be viewed as a type of international law) can be found, for example, in the laws and arrangements

[3] For general reference works on space law see, e.g., Lyall, F., and Larsen, P.B., *Space Law: A Treatise* (2009); Diederiks-Verschoor, I.H.Ph., and Kopal V., *An Introduction to Space Law* (3rd ed. 2008); Christol, C.Q., *Space Law: Past, Present, and Future* (1991); Zhukov, G.P. and Kolosov, Y., *International Space Law* (1984); Jasentuliyana, N. and Lee, R.S.K. (eds.), *Manual on Space Law, Vol. I* (1979); Gál, G., *Space Law* (1969).

of the European Union.[4] Finally, space law exists on the domestic level and can take the form of national laws and regulations – or laws and regulations of territorial subunits, such as the individual states of the United States.[5] This chapter will examine the relationship between the Cape Town Convention and the existing international law of outer space. Domestic space law is treated elsewhere herein to the extent that issues arise in connection with the Cape Town Convention. For example, Section 2.11.14 above discusses the effect of the Convention on domestic licenses for the operation of space assets and the regulations applicable to controlled technology.[6]

Although the need for a separate international law of outer space has been discussed by academics since the early part of the twentieth century, international space law first took shape in the early 1960s in the wake of the launch of Sputnik and other early space activity by the Soviet Union and the United States. The codification of space law began with the *1962 United Nations Declaration of Legal Principles Governing the Activities of States in the Exploration and Use of Outer Space.*[7] This Declaration included the following fundamental principles that formed the bedrock of international space law:

- That space should be explored and used for the benefit of all mankind;[8]
- That space is open to free exploration by all states;[9]
- That no state can claim sovereignty over space or celestial bodies;[10]
- That space should be used in accordance with international law and in the interests of peace;[11]

---

4 Several European states, including some of the current members of the European Union, created the European Space Agency by treaty in 1975. Convention for the Establishment of a European Space Agency (1975), 1981 UKTS 30.

5 Regarding the nature of national regulations that have been adopted by various countries see Jakhu, R.S. (ed.), *National Regulation of Space Activities* (2010).

6 See *supra* Section 2.11.14 regarding licenses and controlled technology.

7 Declaration of Legal Principles Governing the Activities of States in the Exploration and Use of Outer Space, G.A. Res. 1962, 1280th plen. mtg., U.N. Doc. A/RES/1962 (Dec. 13, 1963) [hereinafter Declaration of Principles].

8 Id. para. 1.

9 Id. para. 2.

10 Id. para. 3.

11 Id. para. 4.

- That states must bear responsibility for their activity in space, supervise activities of non-governmental parties, and be liable for harm caused by their activity or by their nationals;[12]
- That states maintain a registry of their space objects;[13]
- That space activity be guided by international cooperation and that states avoid interference with the activities of other states;[14]
- That the state of registry maintain jurisdiction and control over a space object;[15]
- That ownership shall not be affected by an object's entry into space;[16]
- That states provide assistance to astronauts in distress;[17] and
- That states return to the launching state any astronauts or space objects found on their territory or on the high seas.[18]

These principles reflect the issues and concerns that existed at the time of drafting. Space activity was limited to governments and the prospect of private commercial space use was a rather distant concept. The concerns of the United States and the Soviet Union were of a governmental nature, namely, to prevent the appropriation of space by the other state, to avoid the militarization of space, to maintain control over their satellites, and to ensure the rescue and return of their astronauts and spacecraft.

In the years following the adoption of the Declaration of Legal Principles, five treaties were drafted under the auspices of the United Nations to codify, elaborate upon, and expand upon the principles contained in the Declaration. These five core space treaties include:

- The Treaty on Principles Governing the Activities of States in the Exploration and Use of Outer Space, Including the Moon and Other Celestial Bodies (Outer Space Treaty);[19]

---

12 Id. paras. 5 & 8.
13 Id. para. 7.
14 Id. para. 6.
15 Id. para. 7.
16 Id. para. 7.
17 Id. para. 2.
18 Id. paras. 7 & 9.
19 Treaty on Principles Governing the Activities of States in the Exploration and Use of Outer Space, Including the Moon and Other Celestial Bodies, Jan. 27, 1967, 610 U.N.T.S. 205 [hereinafter Outer Space Treaty].

- The Agreement on the Rescue of Astronauts, the Return of Astronauts and the Return of Objects Launched into Space (Rescue and Return Agreement);[20]
- The Convention on International Liability for Damage Caused by Space Objects (Liability Convention);[21]
- The Convention on Registration of Objects Launched into Outer Space (Registration Convention);[22] and
- The Agreement Governing Activities of States on the Moon and Other Celestial Bodies (Moon Agreement).[23]

All of these treaties have entered into force and have been broadly ratified, with the exception of the Moon Agreement which, although it received a sufficient number of ratifications to enter into force, has been ratified by only thirteen states (none of which are space powers).[24] One additional area of international law that could be implicated by the operation of the Cape Town Convention is the international law governing telecommunications. This area of the law regulates the allocation of orbital slots for telecommunications satellites and the use of radio frequencies. This law is overseen by the International Telecommunications Union (ITU) and the governing documents include the *Constitution of the International Telecommunications Union*, the *Convention of the International Telecommunication Union*, and the ITU Radio Regulations.[25] In addition to these core treaties, there are a number of other treaties (primarily bilateral in nature) that do not appear to raise any issues in connection with the

---

20 Agreement on the Rescue of Astronauts, the Return of Astronauts and the Return of Objects Launched into Space, April 22, 1968, 672 U.N.T.S. 119 [hereinafter Rescue and Return Agreement].

21 Convention on International Liability for Damage Caused by Space Objects, Mar. 29, 1972, 961 U.N.T.S. 187 [hereinafter Liability Convention].

22 Convention on Registration of Objects Launched into Outer Space, Jan. 14, 1975, 1023 U.N.T.S. 15 [hereinafter Registration Convention].

23 Agreement Governing Activities of States on the Moon and Other Celestial Bodies, Dec. 18, 1979, 1363 U.N.T.S. 3 [hereinafter Moon Agreement].

24 Status of International Agreements Relating to Activities in Outer Space, Committee on the Peaceful Uses of Outer Space, June 6, 2011, U.N. Doc. A/AC.105/2011/CRP.12.

25 Constitution of the International Telecommunications Union, available at http://www.itu.int/aboutitu/basic-texts/constitution.html [hereinafter ITU Constitution]; Convention of the International Telecommunication Union, available at http://www.ito.int/aboutitu/basic-texts/convention.html [hereinafter ITU Convention].

Cape Town Convention and will not be discussed herein, such as the 1963 *Treaty Banning Nuclear Weapon Tests in the Atmosphere, in Outer Space and Under Water*,[26] instruments that created intergovernmental organizations to develop satellite communications infrastructure,[27] and the intergovernmental agreement between the states involved in the construction and operation of the International Space Station.[28]

In addition to treaty law, space activities must be carried out by states in compliance with customary international law. The recognition of a rule of customary international law requires evidence of the widespread and uniform practice of nations that engage in such practices out of a sense of legal obligation, or *opinio iuris*.[29] One such example of customary international law regarding space activities is found in the customary law requiring that space be used only for peaceful purposes. The Outer Space Treaty only states that "[t]he Moon and other celestial bodies shall be used by all States Parties to the Treaty exclusively for peaceful purposes."[30] A plain reading of this provision suggests that the "peaceful purpose" requirement applies only to use of the Moon and celestial bodies – but does not apply to other regions of outer space, thus allowing for the militarization of the Earth orbits.[31] However, commentators generally agree that customary

---

26 Treaty Banning Nuclear Weapon Tests in the Atmosphere, in Outer Space and Under Water, Aug. 5, 2963, 480 UNTS 43. A complete listing of all space treaties in existence as of 1999 can be found in International Agreements and Other Available Legal Documents Relevant to Space-Related Activities, available at http://www.oosa.unvienna.org/pdf/spacelaw/intlagree.pdf. An updated source of all space treaties and other space-related legal source documents can be found in Dempsey, P. (ed.), *Space Law* (2012).

27 Such organizations include INTELSAT, INMARSAT, EUMETSAT, EUTELSAT, INTERSPUTNIK, and ARABSAT. Lee, R.J., "Reconciling International Space Law with the Commercial Realities of the Twenty-First Century", 4 Singapore Journal of Int'l & Comp. Law 194, 200–201 (2000).

28 International Space Station Intergovernmental Agreement, available at http://www.state.gov/documents/organization/107683.pdf.

29 See Goldsmith, J.L. and Posner, E.A., "A Theory of Customary International Law", 66 Univ. Chicago Law Review 1113, 1116 (1999). In addition to state practice, evidence for the existence of customary international law can also be found in treaties, resolutions, and declarations, court opinions (both domestic and international), and domestic legislation. See Mello, M., "Hagan v. Australia: A Sign of the Emerging Notion of Hate Speech in Customary International Law", 28 Loyola of Los Angeles Int'l & Comp. Law Review 365, 368 (2006).

30 Outer Space Treaty, Art. IV.

31 See Zhukov and Kolosov, *supra* note 3, at 53 (arguing that the Outer Space Treaty requires "total neutralization and demilitarization of celestial bodies and [only] partial

international law (based on long-standing practices of space-faring states) requires that all of outer space, including earth orbits, be used only for "peaceful purposes."[32] Some commentators assert that the fundamental principles of the Outer Space Treaty (including the peaceful use of celestial bodies, the ban on the assertion of sovereignty over celestial bodies, *etc.*) have risen to the level of customary international law and are therefore binding on all states even if a state has not signed the Outer Space Treaty.[33] Although UN General Assembly resolutions are not binding in their own right, they can serve as evidence of customary international law and can therefore exert significant force in governing the activities of states in space. In addition to the 1963 Declaration of General Principles mentioned above, other General Assembly Resolutions have been adopted regarding direct satellite television broadcasting,[34] remote sensing,[35] the use of nuclear power sources in space,[36] and international cooperation in space ventures.[37]

---

demilitarization of outer space"); Bourbonniere, M., "National-Security Law in Outer Space: The Interface of Exploration and Security", 70 J. Air L. & Com. 3, 16 (2005). The position of a number of commentators, as well as the United States government, is that the "peaceful use" requirement does not prohibit all military activity, but only the requires that any military use of space be "non-aggressive," thus permitting, for example, the use of space assets to support military logistics. See, e.g., Bridge, R.L., "International Law and Military Activities in Space," 13 Akron L. Rev. 649, 658 (1979).

32 See, e.g., Bourbonniere, *supra* note 32, at 16; Morgan, R.A., "Military Use of Commercial Communication Satellites: A New Look at the Outer Space Treaty and 'Peaceful Purposes'", 60 J. Air L. & Com. 237, 242 (1994) (explaining that "[s]upport for the conclusion that current international law requires all of outer space to be used for 'peaceful purposes' is compelling.").

33 Lyall and Larsen, *supra* note 3, at 80 (arguing that state practice in combination with UN resolutions, bilateral agreements, and various working arrangements between states make it "difficult to conclude other than that the fundamental principles of the [Outer Space Treaty] are indeed now custom, and as such bind all states.").

34 The 1982 Principles Governing the Use by States of Artificial Earth Satellites for International Direct Television Broadcasting, General Assembly Resolution 37/92 (XXXVII), UN GAOR 37 Supp 51 (UN Doc 1981).

35 The 1986 Principles Relating to Remote Sensing of Earth from Outer Space, General Assembly Resolution 41/65 (XLI), UN GAOR 41 Supp. 53 (UN Doc A/41/53).

36 The 1992 Principles Relevant to the Use of Nuclear Power Sources in Outer Space, General Assembly Resolution 47/68 (XLVII), UN GAOR 47 Supp 49 (UN Doc A/47/610).

37 The 1996 Declaration on International Cooperation in the Exploration and Use of Outer Space for the Benefit and in the Interest of All States, Taking into Account the Needs of Developing Countries, General Assembly Resolution 51/122 (LI).

The Cape Town Convention has ushered in a new era of international space law as the first international treaty that addresses private law, that is, the rights and obligations of parties engaged in business transactions. Earlier international space law applies to commercial space activity in certain respects, but the rights and obligations apply only to states. For example, a state has the duty to supervise the commercial space activities of its nationals.[38] Similarly, the duty of states to return to the launching state errant spacecraft that have crashed in their territory extends to the return of privately owned spacecraft.[39] However, a private company has no standing under international law to demand the return of its errant spacecraft. In contrast, a bank that has an international interest in the form of a security interest in a satellite has a right to exercise remedies under the Convention if the debtor defaults on its payment obligations – because a state that is a party to the Convention is required to enforce the bank's right to exercise these remedies.

The Cape Town Convention also differs from the earlier space treaties in the nature of the concerns that motivated its creation. In contrast to the earlier treaties, the Cape Town Convention is motivated by the concerns that arise from private transactions rather than governmental interests.[40] Rather than being driven by concerns of sovereignty claims and militarization, the Convention addresses the needs of private financiers, such as the priority of secured parties, title to purchased assets, and remedies upon default. This stark difference in the subject matter of the Convention in contrast to the existing space treaties results for the most part in an absence of intersection and conflict between the treaties.[41]

---

38 See *infra* Section 3.4.1.

39 Sundahl, M.J., "The Duty to Rescue Space Tourists and Return Private Spacecraft", 35 J. Space L. 163 (2009).

40 See Larsen, P.B., "Critical Issues in the UNIDROIT Draft Space Protocol", in *Proceedings of the Forty-Fifth Colloquium on the Law of Outer Space* 2, 4 (2003) (explaining that "[t]he Protocol is concerned primarily with private law and with the protections of financiers who enter into private law contracts, whereas existing space law is primarily public law.").

41 van Fenema, P., "The UNIDROIT Space Protocol, the Concept of 'Launching State', Space Traffic Management and the Delimination of Outer Space (Report of the 41st Session of the UNCOPUOS Legal Subcommittee)", 27 Air & Space L. 266, 275 (2002) (explaining that during a meeting in 1997 the UNIDROIT Space Working Group "concluded that there was no conflict between the provisions of the draft Protocol and the existing body of space law" and that the ITU indicated "that it saw neither overlap nor contradiction between

Nevertheless, some intersections do arise and must be kept in mind by practitioners and courts that are involved in the application of the Convention.

The following sections will examine how the Cape Town Convention interacts with space law. However, before taking up these particular points of intersection between existing space law and the Cape Town Convention, the next section discusses the preliminary issue regarding how conflicts between the Convention and other sources of space law should be resolved.

3.2. *The Primacy of the Space Treaties*

The effect of the Cape Town Convention on the existing law of outer space was of critical concern to members of the Space Assets Working Group during the drafting of the Space Assets Protocol. As mentioned above, there was concern among governments and academics that the Space Assets Protocol would inadvertently upset the existing system of international space law. For example, there was concern that the registry of international interests would somehow conflict with the existing register of space objects maintained by the United Nations and negatively affect the legal implications of the UN register. To resolve these concerns the Working Group addressed the relationship of existing space law to the Convention in two

---

the [Cape Town] Convention and the draft Protocol, on the one hand, and the ITU Constitution, Convention and Radio Regulations, on the other."); Larsen, *supra* note 40, at 3 (explaining that the COPUOS Legal Subcommittee "agreed with the no conflicts principle" recognized by the Space Working Group); Lochan, R., "Cape Town Convention & Space Protocol: A Critical Analysis", in Gopalakrishnan, V. and Lochan, R. (eds.), *Conference Proceedings of the ISRO/IISL Space Law Conference 2005: Bringing Space Benefits to the Asian Region*, 6–39, 6–45 (2006) (explaining that "the coexistence and independent functioning of [the Convention and space treaties] may not be a serious cause of concern."); see also Larsen, P.B, "UNIDROIT Space Protocol: Comments on the Relationship between the Protocol and Existing International Space Law, in *Proceedings of the Forty-Fourth Colloquium on the Law of Outer Space* 187, 193 (2002); Ribbelink, O.M., "The Protocol on Matters Specific to Space Assets", European Review of Private Law 37, 41 (2004); Yun, Z., "Revisiting Selected Issues in the Draft Protocol to the Cape Town Convention on Matters Specific to Space Assets", 76 Journal of Air Law & Commerce 805, 827 (2011). For the views of the UNCOPUOS Legal Subcommittee see U.N. Committee of Peaceful Uses of Outer Space, Report of the Legal Subcommittee on its forty-first session, U.N. Doc A/AC.105/787 (2002).

ways in the Protocol. First, the fourth recital in the preamble to the Protocol recognizes the importance of preserving the principles of space law:[42]

> MINDFUL of the established principles of space law, including those contained in the international space treaties of the United Nations and the instruments of the International Telecommunication Union ...

Although the language of a treaty preamble is not binding, it does help to define the context of the treaty which, under the Vienna Convention, must be taken into account when interpreting the treaty.[43] The word "mindful" is not particularly reassuring to those who might wish for a stronger statement of the absolute primacy of existing space law. But the word means at least that the drafters were aware of the importance of existing space law – and perhaps that the treaty should be interpreted to the extent possible in a manner that is consistent with existing space law.

Second, Article XXXV of the Protocol, reproduced here, explicitly establishes the primacy of the UN treaties and ITU instruments over the Convention:[44]

> The Convention as applied to space assets does not affect State Party rights and obligations under the existing United Nations outer space treaties or instruments of the International Telecommunication Union.

This provision is binding on the parties to the Protocol and establishes the primacy of the five United Nations treaties and the ITU instruments over the provisions of the Convention. Although the Protocol is not mentioned in this provision, it would be difficult to deny that the intent of the drafters was not to extend this primacy to the Protocol as well. Unlike the fourth recital of the preamble, only the United Nations treaties and ITU instruments are mentioned in this article, thus leaving any conflicts between the Convention and other sources of international space law to be resolved according to the principles of *lex posterior* and *lex specialis* (which would almost certainly give primacy to the Convention and Protocol). This means that there is at least the potential that some areas of space law will be trumped by the Convention and Protocol. If customary international law or treaties concluded by states outside the auspices of the United Nations

---

42 Space Assets Protocol, Preamble, Fourth Recital.

43 See *supra* Section 2.12.8.

44 Space Assets Protocol, Art. XXXV.

or the ITU conflict with the Convention, primacy will likely be given to the Convention.

To summarize, although the fourth recital of the preamble and Article XXXV of the Protocol appear to establish primacy of existing space law over the Convention and Protocol, this primacy has four potential limitations. First, the recital stating the parties' "mindfulness" of existing space law is something less than a statement of absolute primacy and, in any event, is not binding. Second, Article XXXV does not mention the Protocol, thus opening the door to the argument that while the Convention might trump the existing treaties, the Protocol does not. Third, Article XXXV only grants primacy over the UN treaties and ITU instruments, while failing to mention other sources of space law. Finally, the language of Article XXXV, that the Convention "does not affect State Party rights and obligations under the existing United Nations Outer Space Treaties or instruments of the International Telecommunication Union", might mean something less than the absolute primacy of the Convention. This argument would draw on the fact that Article 45 *bis* of the Convention states that the Convention "prevails" over the *United Nations Convention on the Assignment of Receivables in International Trade*, while in Article XXXIV the Protocol states that the Convention "supersedes" the *Unidroit Convention on International Financial Leasing*.[45] The word "supersedes" is also used in the Aircraft Protocol to establish the primacy of that Protocol over certain other treaties.[46] In the Rolling Stock Protocol, the drafters use the word "prevails" to establish the primacy of that Protocol over two treaties.[47] One might argue that the drafters' choice not to simply state that the Convention "prevails over" or "supersedes" the existing space treaties means that the Convention does not have perfect primacy. On the other hand, the Convention and Protocol use in several places the phrase "does not affect" with the clear meaning that rights under existing law are preserved in their entirety and are not affected by the Convention.[48] It may be that there is no

45 Cape Town Convention, Arts. 45 *bis* & 46.

46 Aircraft Protocol, Arts. XXIII-XXV.

47 Rolling Stock Protocol, Arts. XIX & XX.

48 Cape Town Convention, Arts. 3 & 29(7)(a); Space Assets Protocol, Arts. III, IV(3), XXV(2), XXVI(1), XXVI(2)(c) & XXVII(5). For a general discussion of the relationship of the Convention to its protocols see Chinkin, C., and Kessedjian, C., "The Legal Relationship Between the Proposed UNIDROIT Convention and Its Equipment-specific Protocols", 4 Unif. L. Rev/Revue de droit uniforme 323 (1999).

difference between these various ways of expressing the primacy of one treaty over another, but the use of different language suggests different meanings – and the phrase "does not affect State Party rights and obligations" seems less absolute than saying that the Convention simply "prevails over" or "supersedes" the space treaties.[49] As mentioned above, this issue may be settled in future litigation, but at this point the current language could be said to lack perfect clarity.

3.3. *The Definitions of "Space", "Space Object" and "Space Asset"*

As discussed briefly above, the definition of "space asset" in the Protocol requires (among other things) that the asset either be "in space or designed to be launched into space."[50] This definition, in turn, raises the question of the delimitation of space or, in other words, where does space begin? The Protocol defines "space" as "outer space, including the Moon and other celestial bodies", but this definition does not help to identify the boundary of outer space.[51] Without a further definition in the Protocol of the term "outer space", practitioners and courts must rely on other applicable law when determining whether a particular asset at issue qualifies as a "space asset." The answer to this question will, of course, determine whether the Convention will govern the transaction involving the asset at issue – which will be important for the parties to know when designing a transaction and for a court to know when resolving a dispute regarding the asset. For if an asset will not (and was never intended to) enter outer space, then space law will not apply.

In many cases, the answer to the question of whether an asset is in (or will enter) space will be clear. For example, satellites being sent into orbit will undeniably operate in space and will therefore qualify as space assets. The more difficult question will involve equipment that is operating on the border of outer space. For example, a party (or a court) could in good faith raise the question whether a suborbital space plane that reaches a maximum altitude of 105 kilometers (just above the von Kármán line,

49 Lochan expresses some concern whether the language of Article XXXV is adequate to ensure that the states party to the space treaties will retain all of their obligations under the treaties. Lochan, *supra* note 41, at 48.

50 Space Assets Protocol, Art. I(2)(k).

51 Id. Art. I(2)(j).

which lies approximately 100 kilometers above the Earth and is widely viewed as the limit of outer space) is a space asset.[52] The same question could be raised with respect to a space plane that reaches an altitude of 95 kilometers – just short of the von Kármán line. In connection with this issue, it should be remembered that the Protocol will only apply to a vehicle that is "primarily designed for use in space" and does not apply to a vehicle, such as a suborbital spaceplane, that is designed primarily to travel in air space and is only in space for a brief time.[53] However, the question of the delimitation of space is still relevant if, for example, a space plane was in low-altitude space for an extended period of time, as is being contemplated for the next generation of space planes that will be capable of point-to-point travel.

As just mentioned, when parties to a transaction or a court are faced with the question of whether a particular asset qualifies as a "space asset," they will have to look to applicable law outside of the Protocol. None of the existing space treaties provide any additional useful guidance.[54] Although the Outer Space Treaty clearly contemplates its application to activities on the Moon and other celestial bodies, to objects launched into space, and to objects placed into orbit around the Earth, the treaty does not explicitly define where "outer space" space begins.[55] Both the Liability Convention

---

52 The "von Kármán line," which bears the name of the Hungarian aerospace engineer Theodore von Kármán, is that point at the edge of the Earth's atmosphere "where aerodynamic lift is exceeded by centrifugal force." Hobe, S., Schmidt-Tedd, B. and Schrogl, K. (eds.), *Cologne Commentary on Space Law, Vol. 1* 31 (2009). This line lies approximately 100 kilometers above sea level – although its position is not fixed and can therefore change to some degree depending on fluctuation in air density. At this point, a plane can no longer derive any lift from its wings as it travels through the atmosphere unless it travels at the speed of orbital velocity (at which point it would begin to stay aloft due to the physics of orbital space travel rather than air travel). Id.

53 See *supra* Section 2.3.2.

54 By comparison, although the Chicago Convention does not define air space, Annex VII to the convention defines "aircraft" as "any machine that can derive support in the atmosphere from the reactions of the air other than the reactions of the air against the earth's surface." International Civil Aviation Organization, International Standards: Aircraft Nationality and Registration Marks, Annex 7 to the Convention on International Civil Aviation (5th ed. 2003).

55 See Outer Space Treaty, Arts. I & II (addressing "[t]he exploration and use of outer space, including the Moon and other celestial bodies … ."; Art. IV (prohibiting the placement of weapons of mass destruction "in orbit around the Earth"); Arts. VII & VIII (regarding liability for and jurisdiction over object launched into outer space).

and the Registration Convention define space object as the "component parts of a space object as well as its launch vehicle and parts thereof," but the definition only clarifies that "space object" includes its launch vehicle and components. They do not establish the limit of outer space.[56]

In the absence of any helpful treaty law, there are two remaining possible sources of law. One source of law on the delimitation of space is customary international law, but there is not sufficient state practice (or other evidence of customary law) at this point to support an assertion that customary international law exists regarding the boundary of air space and outer space.[57] The other possible source of law is the applicable domestic law. However, as discussed below, only Australia has enacted legislation that sets a clear demarcation of outer space. Since, outside of Australia, no applicable domestic law is likely to exist and customary international law regarding this issue has not ripened, courts and practitioners may still seek guidance from other sources, such as state practice and the writings of commentators. However, given the lack of clarity on this issue, practitioners are advised to be cautious in any determination they make and to take steps to ensure that their clients' interests are protected regardless of what determination a court may ultimately make on the issue.

Although a customary international law regarding the delimitation of space may be in an early stage of emergence, it has not yet ripened. If a customary international law were to come into being, the evidence suggests that it will define the boundary of space as (or near) the von Kármán line. Evidence of state practice in support of this conclusion can be found in Australian law which suggests that the boundary of outer space lies at 100 kilometers since the law requires a license only for the launch of space objects above that elevation.[58] In addition, South African law establishes the border of outer space by defining outer space as "the space above the surface of the earth from a height of which it is in practice possible to operate an object in an orbit around the earth."[59] This would place the

56 Liability Convention, Art. I(d); Registration Convention, Art. I(d).

57 See *supra* Section 3.1 for a discussion of the nature of customary international law.

58 Space Activities Act of 1998 §8, as amended. See Siemon, N. and Freeland, S., "Regulation of Space Activities in Australia", in Jakhu, R.S. (ed.), *National Regulation of Space Activities* 51 (2010).

59 South African Republic Space Affairs Act, No. 84 of 1993, as amended by the Space Affairs Amendment Act of 1995, Section 1(xv), available at www.saflii.org/za/legis/num_act/saa1993113. See also UNCOPUOS Doc. A/AC/105/865/ADD.1, 20 March 2006. A similar

boundary, under optimal circumstances, in the neighborhood of the von Kármán line – although such a low orbit would be extremely difficult to maintain due to atmospheric drag as demonstrated by the fact that the lowest perigees used by satellites have been approximately 150 kilometers above sea level.[60] Further support for customary international law, in a weaker form since it is only a proposed treaty rather than state practice, may be found in the *Draft Treaty on the Prevention of the Placement of Weapons in Outer Space, the Threat or Use of Force Against Outer Space Objects* submitted in 2008 at the UN Conference on Disarmament by China and Russia to ban weapons in outer space. This draft treaty defines outer space as "space beyond the elevation of approximately 100 km above ocean level of the Earth."[61] Most recently, the Federal Aviation Administration of the United States issued the 2006 Human Space Flight Requirements for Crew and Space Flight Participants pursuant to the Commercial Space Launch Amendments Act of 2004.[62] Although these regulations do not define outer space, they are intended to cover the operations of suborbital space tourism companies, such as Virgin Galactic, that are acquiring licenses under the regulations in order to fly their vehicles to 100 kilometers.[63] State practice supporting an even lower elevation for the limit of outer space can be found in the regulations of the United States Air Force which grant an astronaut rating to Air Force officers who fly only fifty miles above the Earth's surface.[64] Seeking a more comprehensive perspective on

approach is found in the Czech Republic's response to the UNCOPUOS survey regarding the practice of states concerning the delimitation of outer space which stated that "the Czech Republic has respected the customary principle of considering as a space activity the launching of space objects, whose purpose it is to orbit the Earth or otherwise move in outer space." UNCOPUOS Doc. A/AC/105/865/ADD.4, 2 February 2009.

60 Barrett, R.J., "Outer Space and Air Space: The Difficulties in Definition", Air University Review (1973), available at www.airpower.au.af.mil/airchronicles/aureview/1973/May-Jun/barrett.html.

61 Draft Treaty on the Prevention of the Placement of Weapons in Outer Space, the Threat or Use of Force Against Outer Space Objects, Art. I(a), available at www.ln.mid.ru/brp_4.nsf/0/0D6E0C64D34F8CFAC32573EE002D082A.

62 Human Space Flight Requirements for Crew and Space Flight Participants, 14 C.F.R. Parts 401, 415, 431, 435, 440 and 460; Commercial Space Launch Amendments Act of 2004, 49 U.S.C. §§ 70101-70305.

63 Virgin Galactic Brochure, available at www.virgingalactic.com/assets/downloads/Virgin_Galactic_Brochure.pdf.

64 See Air Force Instruction 11-402, Aviation and Parachutist Service, Aeronautical Ratings and Badges, Sept. 27, 2007, para. 2.3.2, available at http://www.e-publishing.af.mil/shared/media/epubs/AFI11-402.pdf.

state practice, the Legal Subcommittee of UNCOPUOS surveyed its member states regarding, among other things, their practice concerning the delimitation of outer space. This survey generated a variety of responses from member states, many of which were compiled and published in a single document by UNCOPUOS.[65] Unfortunately, the responses have not revealed any consensus on the issue.

On a related note, the actions of the United Nations organs (such as General Assembly resolutions) have been accepted as evidence to support the recognition of customary international law. To this point, the Legal Subcommittee of UNCOPUOS has received repeated submissions to resolve the uncertainty regarding the delimitation of outer space. However, virtually no progress has been made on the issue.[66] Despite this inability to reach a consensus at the international level, there appears to have emerged at least a majority position that the boundary between air space and outer space should exist at some point near the von Kármán line.[67] In fact, there have been reports of support among most delegations to UNCOPUOS to fix the boundary between 100 to 120 kilometers above sea level.[68]

Briefly stated, there is some evidence in state legislation, state practice, and the expression of state opinion that supports a customary international law of fixing a boundary of space at approximately 100 kilometers above sea level. However, the amount of evidence is still rather thin and is likely not to amount to the widespread evidence that is required to support the existence of customary law.

---

65 Compilation of Replies Received from Member States to the Questionnaire on Possible Legal Issues with Regard to Aerospace Objects (as contained in U.N. Doc. A/AC.105/635 and Addenda), available at www.oosa.unvienna.org/oosa/SpaceLaw/aero/index.html. Serbia's response deserves notice for defining outer space as "space at a distance of 2 million km or more from the Earth". UNCOPUOS Doc. A/AC/105/865/ADD.1, 20 March 2006.

66 Diederiks-Verschoor and Kopal, *supra* note 3, at 17; van Fenema, *supra* note 41 at 272, 274 (explaining that when the issue of the delimitation of outer space was addressed at the 2002 meeting of the Legal Subcommittee of UNCOPUOS it was recognized that no agreement on this issue had been achieved since the matter was first raised in 1966 and that the subcommittee made no progress on the issue at the meeting.). See Historical summary on the consideration of the question on the definition and delimitation of outer space, COPUOS Doc. A/AC 105/769 (18 January 2002). The most recent meeting of the Legal Subcommittee of COPUOS made little progress on this issue. Report of the Legal Subcommittee on its fifty-first session, U.N. Doc. A/AC.105/1003 (10 April 2012).

67 Hobe et al., *supra* note 52, at 31; Lyall and Larsen, *supra* note 3, at 169.

68 Diederiks-Verschoor and Kopal, *supra* note 3, at 18; U.N. Doc. A/AC-105/C-2/2R 314, para. 2 (April 5, 1979); U.N. Doc. A/AC-105/411 p. 29 (April 8, 1988).

In the absence of customary international law and applicable domestic law defining the limit of outer space (which, as stated above will only be found in Australia), courts and practitioners only have the opinions of commentators to rely on when determining if a particular transaction involves a space asset subject to the Cape Town Convention.[69] Commentators have been debating the question of the delimitation of space for decades in the context of the question of what constitutes a "space object" that is subject to the UN space treaties. Two general approaches to the question of the delimitation of space have emerged since this debate began. One approach, known as the *spatialist* approach, is to identify the border of space on the basis of the natural properties of space.[70] The other approach, known as the *functional* approach, is to classify an object as a space object on the basis of its function as a space-based object as opposed to an aircraft.[71] Since we are only concerned with identifying the boundary of outer

---

69 For a discussion of the debate surrounding the delimitation of space see Lyall and Larsen, *supra* note 3, at 153–173; Diederiks-Verschoor and Kopal, *supra* note 3, at 15–20; Hobe et al., *supra* note 52, at 31–33; Gál, *supra* note 3, at 59–116; Galloway, E., "The Definition of Outer Space", in *Proceedings of the 10th Colloquium of Outer Space* 268 (1968); Kopal, V., "What is 'Outer Space' in Astronautics and Space Law", in *Proceedings of the 10th Colloquium of Outer Space* 275 (1968); Zhukov, G.P., "The Problem of the Definition of Outer Space", in *Proceedings of the 10th Colloquium of Outer Space* 271 (1968); Goedhart, R.F.A., *The Never Ending Dispute: Delimitation of Air Space and Outer Space* (1996); Kopal V., "The Question of Defining Outer Space", 8 Journal of Space Law 154 (1980); Qizhi, H., "The Problem of Definition and Delimitation in Outer Space", 10 Journal of Space Law 157 (1982); Cheng, B., "The Legal Status of Outer Space and Relevant Issues: Delimitation of Outer Space and Definition of Peaceful Use", 11 Journal of Space Law 89 (1983); Fernandez-Brital, O., "UNCOPUOS Legal Subcommittee March-April 1993 Meeting and the Limit between Air Space and Outer Space", in *Proceedings of the 37th Colloquium of Outer Space* 244 (1995); Dodge, M.S., "Sovereignty and the Delimitation of Airspace: A Philosophical and Historical Survey", 35 Journal of Space Law 5 (2009); Gorove, K.M., "Delimitation of Outer Space and the Aerospace Object – Where is the Law", 28 Journal of Space Law 11 (2000); Rosenfield, S.B., "Where Air Space Ends and Outer Space Begins", 7 Journal of Space Law 137 (1979); Gorove, S., "Toward a Clarification of the Term 'Space Object' An International Legal and Policy Imperative?", 21 Journal of Space Law 11 (1993).

70 For discussion of the spatialist approach see, in addition to the works cited *supra* in note 69, Luboš Perek, "Scientific Criteria for the Delimitation of Outer Space", 5 Journal of Space Law 111 (1977).

71 For discussion of the functional approach, in addition to the works cited *supra* in note 69, see Stephen Gorove, "Aerospace – Legal and Policy Issues for Air and Space Law", 25 Journal of Space Law 101 (1997). The functional approach avoids the difficulties that come with the spatialist approach, but brings its own challenges. Rather than focusing on the

space, only the arguments under the spatialist approach are of any value in this inquiry.

The most widely-supported proposal under the spatialist approach has been to draw the line between air space and outer space at the von Kármán line.[72] Among the other possible theories under a spatialist approach are to place the boundary between air space and outer space (1) where there is no longer any atmosphere, (2) where the earth's gravitational pull is balanced by that of another celestial body ("Lagrange points"), (3) where the lowest possible orbital perigee for a satellite is located, or (4) where the ability of a state to exercise sovereignty ends.[73]

On a final note, practitioners and courts must also keep in mind the distinction between a "space asset" under the Cape Town Convention and a "space object" under the space treaties. The definition of a space asset is broader than the term "space object." Most notably, a satellite that has not yet been launched will qualify as a "space asset" under the Protocol, but not as a space object under the UN treaties.[74] As a result, for example, if a creditor seeks a court order permitting the repossession of a warehoused satellite, the creditor can proceed under the jurisdictional provisions of the Cape Town Convention without concern that such provisions will be

---

identification of a physical boundary between air space and outer space, the concept behind of the functional approach is to apply air law to those objects engaged in aviation and apply space law to those objects intended to operate in space. Diederiks-Verschoor and Kopal, *supra* note 3, at 18–20; Lyall and Larsen, *supra* note 3, at 169–170. This calculus would apply regardless of where the object happened to be physically located at the time when a legal claim arose. For example, a satellite being launched into space (along with its launch vehicle) would be subject to space law even if the launch was aborted and the object never reached space. The functional approach faces certain problems when faced with hybrid activities, such as a space plane that may be developed for point-to-point travel and spend significant time in both outer space and airspace.

72 Hobe et al., *supra* note 52, at 31; Diederiks-Verschoor and Kopal, *supra* note 3, at 17; Lyall and Larsen, *supra* note 3, at 167–168.

73 Diederiks-Verschoor and Kopal, *supra* note 3, at 17; Lyall and Larsen, *supra* note 3, at 166–169; Gál, *supra* note 3, at 70–97.

74 Larsen, P.B., and Heilbock, J.A., "UNIDROIT Project on Security Interests: How the Project Affects Space Objects", 64 J. Air L. & Com. 703, 736 (1999) (explaining that "space object" includes "all objects launched by man or attempted to be launched into outer space, such as spacecraft, satellites, and space stations including their components and launch vehicles."); Yun, *supra* note 41, at 815 (space object includes those objects "launched or attempted to be launched into space (and possibly those launched in outer space)"); Larsen, *supra* note 40, at 3; Lochan, *supra* note 41, at 6–55.

trumped by the Outer Space Treaty. On the other hand, where an asset qualifies as both a "space asset" and a "space object", the application of the Cape Town Convention may be affected by the provisions of the space treaties.

3.4. *Liability for Damage Caused by a Space Object*

Although the Cape Town Convention and the Protocol do not address issues of liability, liability issues will undoubtedly be a concern of parties to a transaction under the Convention, such as a lessee who takes control of a satellite that subsequently collides with and causes damage to another satellite. The issues raised in connection with the question of liability caused by space objects are complex and certain tensions in the existing international law regarding liability have emerged during the age of commercial space use. On the one hand, the expanded commercial activity contemplated by the Convention will likely exacerbate these tensions. On the other hand, certain aspects of the Cape Town Convention system may actually improve the operation of the existing liability rules by assisting in the attribution of a harmful act to the responsible parties. Both of these intersections between the Cape Town Convention and the law of liability in space are explored in the following sections.

3.4.1. *An Overview of Liability in Space*

Liability for damage caused by space activities can arise from a variety of sources: domestic law, general international law, and international space law. Under domestic law, a company could be sued by a private party for negligent or reckless actions in the event that damage was caused, for example, by a satellite that has fallen to Earth from orbit.[75] Suing a state under this rubric would likely be difficult due to sovereign immunity. However, international law provides for the liability of states (with the understanding that only other states have standing to bring an action under international law). International law generally imposes responsibility on states for wrongful actions that cause harm. This principle was developed by the International Court of Justice in a series of cases including the

[75] Hobe et al., *supra* note 52, at 144.

Corfu Channel case,[76] the *Military and Paramilitary Activities in and against Nicaragua* case,[77] and the *Gabčíkovo-Nagymaros Project* case.[78] The principles of state responsibility were codified by the International Law Commission and adopted by the UN General Assembly in the form of the *Draft articles on Responsibility of States for Internationally Wrongful Acts* (hereinafter the Draft Articles on State Responsibility).[79] In short, a state must pay reparations for any injuries "caused" by an "internationally wrongful act."[80] However, the standard of an "internationally wrongful act" is not easily met since it requires, first, that the act be attributable to a particular state and, second, that the act constitute a breach of an international obligation.[81] Not all actions, even if harmful, constitute a violation of international law. Moreover, the added requirement that the injured state prove that the wrongful act "caused" the injury can in many cases pose significant difficulties.[82]

State responsibility and liability under international law for damage caused by space activity was given greater clarity in the space treaties. The fundamental treaty language establishing state responsibility is set forth in Article VI of the Outer Space Treaty (reproduced here in pertinent part):[83]

> States Parties to the Treaty shall bear international responsibility for national activities in outer space, including the Moon and other celestial bodies, whether such activities are carried on by governmental agencies or by non-governmental entities, and for assuring that national activities are carried out in conformity with the provisions set forth in the present Treaty. The activities of non-governmental entities in outer space, including the Moon and other celestial bodies, shall require authorization and continuing supervision by the appropriate State Party to the Treaty.

[76] *Corfu Channel, Merits, Judgment*, I.C.J. Reports 1949, p. 4, at p. 23.

[77] *Military and Paramilitary Activities in and against Nicaragua* (*Nicaragua v. United States of America*), Merits, Judgment, I.C.J. Reports 1986, p. 14, at p. 142, para. 283, and p. 149, para. 292.

[78] *Gabčíkovo-Nagymaros Project* (*Hungary v Slovakia*), I.C.J. Reports 1997 at p. 38, para. 47.

[79] See Draft Articles on Responsibility of States for Internationally Wrongful Acts, with Commentaries, available at http://untreaty.un.org/ilc/texts/instruments/english/commentaries/9_6_2001.pdf [hereinafter Draft Articles]. See also Larsen p. 103–104.

[80] Draft Articles, Art. 31(1).

[81] Id. Art. 2.

[82] Id. Art. 31 (Comm. 10).

[83] Outer Space Treaty, Art. VI.

Beginning with the first sentence of this article, a state is required to "bear international responsibility for national activities in outer space ... whether such activities are carried on by governmental agencies or by non-governmental entities." Although there is some difference of opinion regarding the precise meaning of this provision, it can be taken in the broadest sense as imposing responsibility upon State A for any violation of international law resulting from space activity undertaken by (1) the government of State A or by any private entity acting on behalf of State A and (2) any non-governmental entities (including private companies) over whose activity the state has jurisdiction (whether this jurisdiction arises from the basis of nationality, territoriality, the registration of the space object, or any other basis under international law).[84] In one case, the Netherlands adopted this broader interpretation of its responsibility by accepting Article VI responsibility for the activities of New Skies NV following the transfer of INTELSAT satellites to this Dutch entity, even though it was not the launching state or state of registry.[85]

This primary statement of state responsibility is either clarified (as some see it) or supplemented (as others see it) by the second sentence of Article VI, reproduced above, which requires the "appropriate State" to authorize and continually supervise all space activity by non-governmental entities, such as private companies.[86] Some interpret this sentence as only clarifying the responsibility already imposed under the first sentence by making explicit that this responsibility entails the duty to authorize and supervise its national activity.[87] Others see it as a separate imposition of

---

84 Hobe et al., *supra* note 52, at 111–114; von der Dunk, F.G., "Liability versus Responsibility in Space Law: Misconception or Misconstruction?", in *Proceedings of the Thirty-Fourth Colloquium on the Law of Outer Space* 363, 367 (1991) (arguing that the "national activities must be deemed to encompass *both* activities undertaken by nationals, whether natural or juridical, *and* activities undertaken from within the territory of the state in question."); Ribbelink, *supra* note 41, at 42; Chatterjee, J., "Reflections on the Concepts of 'Liability' and 'Jurisdiction' in a Privatised Atmosphere, in Sundahl, M. and Gopalaskrishnan, V. (eds.), *New Perspectives on Space Law* at 183, 189–191 (2011); see also von der Dunk, F.G., "Public Space Law and Private Enterprise – The Fitness of International Space Law Instruments for Private Space Activities", in *Legal Framework for Privatising Space Activities, Proceedings of the Project 2001 Workshop on Legal Issues of Privatising Space Activities* at 12 (1999).

85 Lyall and Larsen, *supra* note 3, at 92 (citing Note Verbale of 29 July 2003 from The Netherlands, U.N. Doc. A/AC105/806).

86 Outer Space Treaty, Art. VI.

87 Hobe et al., *supra* note 52, at 111.

responsibility on a new category of states that qualify as the "appropriate states."[88] For those scholars espousing this latter point of view, the definition of "appropriate state" becomes an issue of contention. Some scholars have posited that only a "launching state" or the state of nationality of the actor should qualify as an "appropriate state."[89] However, others have argued that the term "appropriate state" must, due to its vague nature, be interpreted broadly to potentially include any state that has jurisdiction over the particular space activity (however such jurisdiction arises) or is substantially involved in the operation or control of the space object.[90]

If the second sentence of Article VI is viewed as merely clarifying the responsibility imposed by the first sentence, then state responsibility for failure to properly authorize or supervise space activity would arise from the responsibility explicitly imposed by the Article's first sentence. On the other hand, if the duty to authorize and supervise is seen as a separate duty imposed on an "appropriate state", state responsibility for the failure to carry out this duty would spring from general principles of international law as codified in the above-mentioned Draft Articles on State Responsibility.[91] The "internationally wrongful act" required under the Draft Articles on State Responsibility would be the breach of the duty of the "appropriate state" to authorize and supervise the space activity.

Since Article VI uses the term "responsibility" rather than "liability" (which appears in Article VII), the difference between these terms should

88 For a thorough discussion of this issue see Chatterjee, *supra* note 84, at 194–197.

89 See, e.g., Gorove, S., "Liability in Space Law: An Overview", 8 Ann. Air & Space L. 377 (1983); Vladlen Vereshchetin, V., "Space Activities of 'Nongovernmental Entities': Issues of International and Domestic Law", in *Proceedings of the 26th Colloquium on the Law of Outer Space* at 261 (1983).

90 See, e.g., Wirin, W.B., "Practical Implications of Launching State – Appropriate State Definitions", in *Proceedings of the 37th Colloquium on the Law of Outer Space* at 113 (1994); Kerrest, A., "Remarks on the Responsibility and Liability for Damage Caused by Private Activity in Outer Space", in *Proceedings of the 42nd Colloquium on the Law of Outer Space* at 139 (1997); Bourély, M., "Rules of International Law Governing the Commercialisation of Space Activities", in *Proceedings of the 29th Colloquium on the Law of Outer Space* at 157 (1986); Gerhard, M., "Transfer of Operation and Control with Respect to Space Objects – Problems with Responsibility and Liability of States", 51 Zeitschrift für Luft- und Weltraumrecht 571, 571(2002); Hobe et al., *supra* note 52, at 117.

91 Regarding liability arising in this instance from the first sentence of Article VI see Hobe et al., *supra* note 52, at 110–111, 116.

be considered.[92] Although responsibility and liability are similar concepts and are often used interchangeably, there is a distinction between the concepts under international law. In short, the concept of responsibility is broader than the concept of liability. While liability means that the liable state must pay reparations for damages caused by the state, responsibility entails additional obligations. Bin Cheng has captured this broader sense of the concept of responsibility by explaining that a responsible state is "answerable" for a wrongful act – and not merely liable.[93] Under the Draft Articles on State Responsibility, a state that is found to be responsible for a wrongful act has, in addition to the duty to pay reparations (which may take the form of restitution, compensation and/or satisfaction),[94] the duty to (1) continue performance of its international obligations,[95] (2) cease the wrongful conduct,[96] and (3) provide, under certain circumstances, assurances of non-repetition of the wrongful conduct.[97]

Another significant difference between the liability imposed on the "launching state" by Article VII of the Outer Space Treaty (discussed below) and the international responsibility imposed by Article VI is that the state that is found to be responsible need only pay reparations if it has committed an "internationally wrongful act", whereas liability imposed on the launching state can result even if the conduct causing the damage was lawful.[98]

In addition to the state responsibility imposed by Article VI, the existing space law treaties place liability for damage caused by space objects on those states involved with the launch of the object. Article VII of the Outer Space Treaty states the following with respect to this liability:[99]

---

92 Regarding the difference between state responsibility and liability see von der Dunk, *supra* note 84, at 364–365.

93 Cheng, B., "Article VI of the 1967 Space Treaty Revisited: 'International Responsibility', 'National Activities' and 'The Appropriate State'", 26 J. of Space L. 7, 9 (1998).

94 Draft Articles, Arts. 31 & 34.

95 Id. Art. 29.

96 Id. Art. 30.

97 Id.

98 The Draft Articles explain that the Liability Convention "concerns liability for lawful conduct rather than responsibility in the sense" described in the Draft Articles. Id. at 125. See also id. at 31.

99 Outer Space Treaty, Art. VII.

> Each State Party to the Treaty that launches or procures the launching of an object into outer space, including the Moon and other celestial bodies, and each State Party from whose territory or facility an object is launched, is internationally liable for damage to another State Party to the Treaty or to its natural or juridical persons by such object or its component parts on the Earth, in air space or in outer space, including the Moon and other celestial bodies.

This provision assigns liability to a variety of actors that include (1) the state that conducts the launch, (2) the state that procures the launch, (3) the state from which an object is launched, and (4) the state that owns the facility from where the object is launched. This liability extends to damage caused by an object or its component parts (in the event of a breakup) and applies to damage occurring anywhere: on Earth, in the air, or in space.

The 1972 Liability Convention elaborates on the nature of a state's liability for damage caused by space objects and addresses certain issues that were left open in the Outer Space Treaty, such as the standards for finding liability and the process for seeking a remedy. The general liability imposed by Article VII of the Outer Space Treaty is divided into two categories by the Liability Convention. Article II of the Liability Convention imposes strict liability on the "launching State" for all damage caused by a space object on Earth or to aircraft in flight:[100]

> A launching State shall be absolutely liable to pay compensation for damage caused by its space object on the surface of the Earth or to aircraft in flight.

In contrast, Article III of the Convention imposes liability on the launching state when a space object causes damage to the space object of another state when the object is in the air or in space, but only when the launching state is "at fault":[101]

> In the event of damage being caused elsewhere than on the surface of the Earth to a space object of one launching State or to persons or property on board such a space object by a space object of another launching State, the latter shall be liable only if the damage is due to its fault or the fault of persons for whom it is responsible.

---

100 Liability Convention, Art. II. "Space object" is broadly defined in Article I of the Liability Convention as including "component parts of a space object as well as its launch vehicle and parts thereof." Id. Art. I(d).

101 Id. Art. III. Note that both Article II and Article III address damage caused in airspace, with the distinction being that Article II governs damage caused to aircraft while Article III addresses damage to space objects (which are presumably either *en route* to or returning from space).

One troubling aspect of Article III is that it lacks a clear standard for determining when a state (or a person for whom a state is responsible) would be found at "fault" for causing damage. Would a showing of intent be necessary? Or would recklessness, gross negligence, or simple negligence be sufficient?[102] Liability is also complicated (and this applies not only to situations governed by Article III) by the difficulty in attributing damage to a particular state, particularly when damage is caused by unidentifiable component of a space object.[103]

Under Article I of the Liability Convention, the definition of a "launching State" that could bear liability for damage caused by a space object remains unchanged from the Outer Space Treaty and includes (1) a state which launches a space object, (2) a state which procures the launch of a space object, (3) a state from whose territory a space object is launched and (4) a state from whose facility a space object is launched.[104] This last category could apply to a state that owns a launching facility, for example, located on an artificial island on the high seas. In the event that there are multiple launching states, each state is jointly and severally liable for damage caused by the space object.[105] This multi-faceted definition of a launching state together with the rule of joint and several liability provides a state that has suffered damage (or whose nationals have suffered damage) the possibility of pursuing multiple states, thus allowing for a greater likelihood of recovery.[106]

---

[102] For a discussion of the possible standards that could be applied to determine fault see Hurwitz, B.A., *State Liability for Outer Space Activities* 32–36 (1992); Forkosch, M.D., *Outer Space Liability* 80–81 (1982).

[103] The problem of identification is particularly challenging when damage is caused by space debris. See Sundahl, M.J., "Unidentified Orbital Debris: The Case for a Market-Share Liability Regime", 24 Hastings Int'l & Comp. Law Review 125, 136 (2000).

[104] Liability Convention, Art. I.

[105] Id. Art. V(1) & (3).

[106] See Hobe et al., *supra* note 52, at 135; see also Lee, *supra* note 27, at 200–201 (explaining how there could, theoretically, "be well over four launching States in any single launch."). Liability also attaches if an attempted, but unsuccessful, launch of a space object causes harm. Liability Convention Art. I(b). The Liability Convention addresses additional related issues that are not directly germane to the discussion at hand including, but not limited to (1) liability in the event that objects of two states collide and cause damage to a third state, (2) joint and several liability for joint ventures, (3) exoneration from liability, (4) limitations on liability, and (5) the claims process. For discussion of these and other issues related to liability see the works listed *supra* in note 3. See also Hurwitz, *supra* note 102 and Forkosch, *supra* note 102.

3.4.2. *How the Protocol May Increase Existing Tensions in the Law Governing Liability in Space*

The paradigm for state liability for space activities established by the Outer Space Treaty and the Liability Convention has been placed under stress over the years by the gradual expansion of private commercial activity in space. The liability paradigm set forth in the treaties primarily contemplates governmental space activity which involves a state launching a satellite or spacecraft and maintaining control over that space object for its entire lifespan.[107] With the advent and expansion of private space activity, the complexities of commercial transactions, including the transfer of ownership of a space object during its operational lifespan (such as on-orbit satellites), challenged the propriety of the existing liability paradigm. Some of the points of stress in this paradigm that have begun to appear are illustrated in the scenarios described in following paragraphs. While these issues have existed since the commercialization of space began, some of the tensions between commercialization and the existing law of outer space will be exacerbated by the use of the Cape Town Convention.

One aspect of existing space law that is likely to face greater scrutiny as we move into a time of increasing commercialization of space is whether it is appropriate and desirable for the launching state to be liable for damage caused by space objects that are launched and controlled by private entities.[108] Although imposing liability on the launching state has the effect of motivating the launching state to take greater care in the supervision of private space activity, at some point the sheer volume of commercial space activity may render this potential liability a detrimental burden. One consequence of continuing to impose this liability on the launching state is likely to be an unreasonable burden on the space industry that crushes innovation. One example of how this burden of potential liability can burden a private entity can be found in the requirement of national space laws

---

[107] Although governmental use of space was the primary scenario contemplated by the drafters of the space treaties, there was also an understanding that commercial activity would also take place, as is made clear in the Outer Space Treaty's reference in Article VI to space activity by "non-governmental entities."

[108] von der Dunk, F.G., "The Illogical Link: Launching, Liability and Leasing", in *Proceedings of the Thirty-Sixth Colloquium on the Law of Outer Space* at 349, 351 (1993) (noting that "[p]rivate entities themselves can never qualify as 'launching states' and therefore cannot incur liability by means of international space law.").

that private entities show financial responsibility, by way of obtaining insurance or otherwise to compensate third parties who suffer injury as a result of a launch.[109] Although such excessive regulation cannot yet be said to have occurred, it is a potential result that should be considered as space law evolves in the new era of commercial space use.

Another widely criticized shortcoming of the existing law of outer space is that the launching state continues to be liable under the Outer Space Treaty and Liability Convention even after the transfer of the space object by the launching state, or by one of its nationals, to another state or private party (or after the state's loss of direct control over the object through other circumstances, such as the lease of an object).[110] The liability imposed by the treaties on a launching state is perpetual, with only a few narrow exceptions.[111] For example, if State X launches a satellite, sells the satellite five

---

109 See, e.g., the United States requirements for financial responsibility of private entities engaged in a launch. 140 C.F.R. Part 440. Of course, the requirement of financial responsibility provides a benefit by ensuring that innocent private parties who are damaged by private space activity and have no recourse under the Liability Convention will be made whole.

110 See, e.g., von der Dunk, *supra* note 108, at 351; Chatzipanagiotis, M., "Registration of Space Objects and Transfer of Ownership in Orbit", 56 Zeitschrift für Luft- und Weltraumrecht 229, 230 (2007); Larsen, *supra* note 40, at 5; Gerhard, *supra* note 90, at 575; Hertzfeld, H.R., and von der Dunk, F.G., "Bringing Space Law into the Commercial World: Property Rights Without Sovereignty", 6 Chicago Journal of Int'l Law 81, 89 (2005); Ribbelink, *supra* note 41, at 42–43; Lee, R.J., "Effects of Satellite Ownership Transfers on the Liability of the Launching States", in *Proceedings of the Forty-Third Colloquium on the Law of Outer Space* 148, 151 (2001). 151; Larsen, *supra* note 41, at 190; U.N. Doc. A/AC.105/C.2/L.225, 23 Jan. 2001. The transfer of control over a space object can also result from the sale of a company that owns the object. Although the company's nationality would not likely change upon its sale, since the place of incorporation would probably remain unchanged, the control of company affairs could pass to foreign shareholders. Lee, *supra* note 110, at 151–152. An example of the lease of a satellite occurred in 2002 when the company SES S.A., a Luxembourg company, leased the Astra-1B satellite to the Swedish company Nordic Satellite AB. See Gerhard, *supra* note 90, at 573. For additional examples of satellite leasings see von der Dunk, *supra* note 108. An example of a private party acquiring control of a satellite from an original owner in another way can be found in the taking of title by Lloyds of the Indonesian Palapa B satellite following the payment of insurance proceeds to the insured. Larsen, *supra* note 41, at 190.

111 Exoneration from liability is possible under Article VI of the Liability Convention if the damage was caused by the gross negligence (or the intentionally harmful act or omission) of the claimant state. Liability Convention, Art. VI(1). A state is also free from liability in the event that the damage is incurred by its own nationals or by the nationals of

years later to State Y, and, due to State Y's negligence, the satellite reenters the atmosphere and causes extensive damage on the surface of the Earth, State X will bear absolute liability for the damage under the Liability Convention.[112] A similar scenario would arise in the event that the launching state leases a satellite to a third party and damage on Earth results from the lessee's negligence. Critics of this result see a fundamental unfairness in imposing liability on the launching state when it was in no way involved in the harmful activity.[113] The free trade of satellites and other space objects may be inhibited due to the liability that launching states face if the object causes harm while under the control of the buyer or lessee. For damage caused by the sold satellite in space, the problem is not as likely to arise since liability for damage in space requires the showing of the launching state's fault. Such fault is unlikely to be found if the launching state had no control over the object or the operator at the time that the damage occurred.[114]

---

another state who are "participating in the operation" of the space object or are in the "immediate vicinity" of the launching or recovery of the object at the invitation of the launching state. Id. Art. VII.

112 State X may not be liable in this scenario if the damage occurred within the territory of State Y pursuant to the exoneration provision of Article VI of the Liability Convention. Id. Art. VI(1). The analysis is in some ways different if a satellite is transferred by a launching state (or its national) to a state that was also involved in the launch of the object and is therefore itself a "launching state" (or to a national of a launching state). In this situation, while the transferor would continue to be liable as a launching state, the transferee state would also be liable as a launching state. Although the same type of risk arises (that the object would cause damage on the surface of the Earth or an aircraft in flight while under the control of the transferee), the risk is somewhat reduced since the transferee state would be likely to exercise greater care in the operation of the space object since it is itself liable as a launching state.

113 It should be noted that the imposition of liability on a launching state after it has relinquished control of a space object is not an issue that arises only upon the transfer or lease of an object. For example, if State A launches a satellite from the territory of State B and State A maintains exclusive control over the satellite, State B will still face absolute liability as a "launching state" for any damage caused by the satellite to an aircraft or on the surface of the Earth even if it never had any control over the object. Some commentators have proposed a new liability regime that would replace liability placed solely, and immutably, on the launching state. See, e.g., Lee, *supra* note 110, at 148 (proposing that liability rest with the state (or entity) that has actual control over the space object).

114 If the launching state is not liable for damage in space on the grounds that it was not at fault because it was not in control of the object, this does not mean that no party can be

This risk of a launching state being held liable for damage caused after the sale or lease of the satellite or other space object can be minimized by restrictions on the transfer of the object under domestic law. If the state is the owner of the object, it can simply decide not to transfer the object. The problem is more complicated if a private entity subject to the state's jurisdiction owns the object. The state can enact legislation prohibiting its nationals from transferring a space object without state permission – a type of legislation that is in effect in many countries. The state can also require its nationals to obtain insurance and reimburse the state for any losses incurred.[115] States and private entities can also enter into indemnification agreements upon the transfer of space assets in order to recoup any amounts that they are required to pay out in the event of an accident. However, such provisions have limited value if, for example, the buyer is a private party with limited financial resources that is unable to fully cover the costs incurred by the launching state.[116] A launching state could also enter into a bilateral agreement with the transferee's state of nationality pursuant to which the transferee's state agrees to indemnify the launching state for any costs incurred as a result of the transferee's actions.[117] However, there may be little impetus for the transferee's state to enter into such an agreement.

A further issue related to liability that may attract increased debate as the Cape Town Convention comes into use is the issue of the scope of the

found liable. If the transferee with control over the object is a state, then it will bear international responsibility under Article VI of the Outer Space Treaty and, under a broad interpretation of Article VI, a state of nationality of a private transferee will likewise bear responsibility for damage caused by the object. Lyall and Larsen, *supra* note 3, at 104–105; Smith, L.J., and Baumann, I. (eds.), *Contracting for Space* at 52 (2011); Lee, *supra* note 110, at 151–152; Jasentuliyana and Lee, *supra* note 3, at 17. However, as explained above, responsibility under Article VI will only arise if a violation of international law arises. This rules out a finding of liability for lawful conduct unless a successful claim can be brought under domestic law, such as tort law. A violation of international law leading to state liability could arise if a state fails to supervise the space activity of its nationals (as required under Article VI of the Outer Space Treaty) and such failure to supervise results in damage. See Gerhard, *supra* note 90, at 576 and note 18.

115 See Lee, *supra* note 110, at 149 (discussing financial responsibility requirements under U.S. and Australian law); Gerhard, *supra* note 90, at 577.

116 Hertzfeld and von der Dunk, *supra* note 110, at 89–90.

117 See Gerhard, *supra* note 90, at 576; Ribbelink, *supra* note 41, at 44; Chatzipanagiotis, *supra* note 110, at 233–234.

definition of "launching state." Two questions in particular touch upon the increased commercialization of space. First, is the state of incorporation of a private company always a launching state? And is this true even when the private launch takes place from the territory of another state or from a location outside the territory of any state? The latter scenario reflects the practice of Sea Launch AG, a Swiss corporation that conducts launches from a converted oil drilling platform in the high seas of the Pacific Ocean. Expert opinion weighs in favor of treating the state of nationality of a company as a launching state on the basis that the activity of a state's nationals should be imputed to the state (a conclusion that can be drawn from Article VI of the Outer Space Treaty which makes a state responsible for the activity of its nationals).[118] But it has also been suggested that such launches from international territory do not so easily allow for the identification of a launching state and that there is a lacuna in the law with respect to this issue.[119] If a state happens to register with the UN an object launched by a private company, then this is a clear indication that the state is a launching state since, under Article II of the Registration Convention, only a launching state can register a space object.

The second question concerning the scope of the definition of "launching state" more directly implicates the Cape Town Convention and has a less certain answer. The question is whether the provision of financing by a state (or by a state's national) will be viewed as the state's "procurement" of the launch and will therefore result in that state being treated as a "launching state." Some commentators have taken a broad view of the concept of "procurement" by finding that it embraces any kind of assistance in the launching of a space object, including the provision of any economic benefits.[120] If this would include a state's funding of a launch, would the provision of financing by a private bank make its state of nationality a "launching state" as well? One could reasonably argue that this type of private activity

---

118 See Kerrest, A., "Launching Spacecraft from the Sea and the Outer Space Treaty: The Sea Launch Project," in *Proceedings of the 40th Colloquium on the Law of Outer Space* 264, 267 (1997).

119 See Schrogl, K. and Davies, C., "A New Look at the 'Launching State": The Results of the UNCOPUOS Legal Subcommittee Working Group 'Review of the Concept of the Launching State 2000–2002'," in *Proceedings of the 45th Colloquium on the Law of Outer Space* 286, 291 (2002).

120 Wirin, *supra* note 90, at 111.

which makes a launch possible should be imputed to the entity's state of nationality just as the direct physical involvement of a private entity in a launch could be imputed to the state.

The aforementioned issues – namely state liability for private activity, the continued liability of launching states for damage caused after relinquishing control through sale or lease, and the treatment of states that provide financing as "launching states" – have been in existence from the advent of commercial space use.[121] However, these issues will be exacerbated by the entry of the Cape Town Convention into force with respect to space assets under the theory that the Convention will result in an increase in the sale and lease of space assets after their launch. The increase in sales and leases under the Convention will arise not only in simple sales and leases, but through sales and leases that take place in the form of remedies pursued by secured creditors. This increase in transactions is likely to occur if the Cape Town Convention is perceived as being useful to facilitate such transactions by providing rights and remedies, or other benefits, which were not available under pre-existing law. The increase in sales and leases will create more situations in which the launching state loses control over the space object (or loses control, by way of loss of jurisdiction, over the entity that controls the space object), thus making more distinct the disconnection between the existing liability regime and the realities of commercial space activity. In addition, the right of a secured creditor to repossess a space asset or take control of the asset upon default (in advance of selling or leasing the asset) will place additional stress on the existing liability regime by potentially interfering with the launching state's ability to control the asset even before it is sold or leased. For the launching state will potentially bear liability for any damage caused by the asset (and will bear absolute liability for any damage caused in the air or on Earth) even though the launching state has been deprived of any ability to control the asset due to secured creditors seizing of control under the Cape Town Convention.

---

[121] See, e.g., van Fenema, *supra* note 41, at 275 (explaining that the Legal Subcommittee of UNCOPUOS realized that the problem presented by the perpetual liability of the launching state "arose independently from the [Cape Town] Convention and Protocol, as a result of private enterprise activities and related transfers of ownership.").

3.4.3. *How the Convention May Assist in Determining Liability*

In contrast to the foregoing discussion of how the Cape Town Convention could worsen existing flaws in the law of liability in space, the Cape Town Convention may also turn out to have a beneficial effect by facilitating the operation of the liability provisions of the space treaties in certain ways because the registry of international interests will (1) aid in the identification of space objects that have caused damage and (2) aid in the identification of entities that have actual control over space objects. The identification of the private party in control of the space asset that has caused harm will aid in bringing an action for damages against that party under applicable domestic law. Moreover, the registry of international interests may aid in identifying the state that bears liability under international law.

As mentioned above, one of the challenges of finding a state liable for damage caused by a space object is identifying the launching state. The United Nations register of space objects that is maintained pursuant to the Registration Convention plays a critical role in helping to identify launching states. The Registration Convention was adopted in 1975 to establish a central international register of space objects for a variety of reasons including (1) improving awareness of objects in orbit, (2) establishing the jurisdiction and control over space objects by the state of registry, and (3) identifying the launching state of objects that cause damage.[122] Pursuant to Article IV of the Registration Convention, the launching state (or one of the launching states if there are more than one) must submit the following information to the UN Secretary-General for inclusion in the international register upon the launching of a space object:[123]

- Name of the launching state (or launching states)
- An appropriate designator of the space object or its registration number
- Date and territory or location of launch
- Basic orbital parameters
- General function of the space object

122 Regarding the Registration Convention see, generally, Lyall and Larsen, *supra* note 3, at 84–96.

123 Registration Convention, Art. IV(1). The launching state is also required to maintain a domestic registry of its space objects. Such domestic registries need not be accessible to the public and the information contained in the registries is left to the discretion of each state. Id. Art. II(1) & (3). An additional international registry of all launches is

The United Nations maintains a searchable "Online Index of Objects Launched into Outer Space" which includes space objects that have been registered pursuant to the Registration Convention and others that have not.[124] Launching states can amend their registration with additional information and must also notify the Secretary-General if an object is no longer in orbit.[125] In the event that damage is caused by a space object, the register enables the damaged party to identify the launching states that may be subject to liability under the Liability Convention. This presumes, of course, that the space object can somehow be identified, whether by its markings, its location in orbit, or some other means. This task of identification is complicated by the highly destructive nature of collisions involving space objects.

The Cape Town Convention registry of international interests in space assets can serve the same function as the UN register by assisting in the identification of the launching state or other states (or entities) that may face liability for damage caused by a space object.[126] But rather than merely duplicating the benefits of the UN register, the registry of international interests will supplement the UN register by providing information that may not be available through the UN register. Like the information collected under the Registration Convention, the registry of international interests will be available to the public online. However, the type of information provided on the registry will differ significantly from information found on the UN register. The registry of international interests will be searchable by asset and the records returned upon a search will provide the name of any parties that have registered or recorded any interest in the asset. In addition, instead of providing the names of the launching states, the registry of international interests will contain the names of the holders of an international interest in the asset or a buyer that has registered a sale. The registry of international interests also differs from the UN register in that (1) states register in the UN register to comply with international law while parties register international interests to protect their financial

maintained by the UN Committee on the Peaceful Uses of Outer Space pursuant to UN General Assembly Resolution 1721 (XVI) 1961. The information collected pursuant to this resolution is made available by the UN in the A/AC.105/INF document series.

124 *Online Index of Objects Launched into Outer Space* can be accessed at http://www.unoosa.org/oosa/showSearch.do.

125 Registration Convention, Art. IV(2) & (3).

126 This possibility is raised in Larsen, *supra* note 40, at 6; see also Yun, *supra* note 41, at 828.

interests and (2) states are only required to submit information to the UN when an object has reached space while holders of an international interest are likely to register prospectively. In light of these differences between the two registries, the Cape Town Convention registry could complement the UN register in a number of ways.

First, a space object that is not on the UN register may be registered on the Cape Town Convention registry. If the launching state is not a party to the Registration Convention, but is involved in a transaction related to a space object that gives rise to a registrable interest under the Space Assets Protocol, there will likely be a registration related to the object on the Cape Town Convention registry.[127] A party that suffers damage from the object could then use the registry to identify the launching state. For example, if State A purchases a satellite which later falls from orbit and causes damage in State B, State B will be able to identify State A as the owner of the satellite (and, in turn, as the launching state) by searching the registry of international interests. An object may also be registered under the Cape Town Convention, but not yet on the UN register, when the object has not yet entered space. To illustrate this point, consider the situation in which a private company incorporated in State X procures the launch of a satellite which fails to reach a sustainable orbit and causes damage in State Y after reentering the atmosphere soon after launch. If the satellite were financed through a secured loan from Bank One, the bank would have likely registered this security interest long before the launch took place – while State X may not have submitted any information about the launch to the UN Secretary-General. After discovering that Bank One held a security interest in the satellite, the damaged party could approach Bank One to discover the identity of the debtor who launched the satellite. An action could then be brought against the private entity that owned the satellite or against State X as the launching state. This identifying information could also be used by a state that is required to pay damages under the Liability Convention as the launching state and, in turn, desires to seek recourse against the private operator that was in actual control of the space object that caused damage.

---

[127] Larsen, *supra* note 41, at 188 (explaining that while the Registration Convention requires registration of space objects by states, the Cape Town Convention registrations will be made by private parties to a financial transaction).

Second, a space object that causes damage on Earth may be easier to identify through information that is found on the Cape Town Convention registry and may not be on the UN register, such as a serial number.[128] The registry of international interests will also carry information about components of space objects (since international interests in components are separately registrable), which could assist in the identification of the space object as a whole. Although a search for international interests will only reveal the name of the buyer or creditor, this information can lead, through further inquiry, to the identity of the launching state or other responsible party.

Third, the Cape Town Convention registry will capture transfers of ownership of space objects (and the transfer of control through leases) through the registration of sales and leases. This information could be useful if a damaged state is seeking reparations under the theory that the "appropriate state" is responsible for the damage due to its failure to properly authorize or supervise the space activity that caused the damage. For example, consider a scenario in which State X places a satellite in orbit, the satellite is subsequently sold to a private company in State Y, and damage is caused to State Z as a result of State Y's failure to supervise the operation of the satellite. The UN register would not be helpful in identifying State Y as the responsible party. However, if the new owner of the satellite registered its purchase of the satellite under the Space Assets Protocol, a search of the registry would reveal the identity of the owner – and would support the contention that State Y, as the owner's state of nationality, was responsible for supervising the owner's space activity. Similarly, the registration of sales and leases will be helpful to determine what party had actual control of a space object at the time damage occurred, which could help support a claim under state responsibility for its national activities in space or under domestic tort laws. For example, if a space station is launched by State A, is subsequently leased to State B, and then collides with State C's satellite due to State B's negligent operation of the space station, State C would be able to determine that State B was in control of the space station by searching the Cape Town Convention registry (where it would find the registration of the lease). In a final example, damage could be caused by a space object while under the control of a secured party that had taken over control of

128 See, e.g., id. (explaining that the Protocol has more stringent identification requirements than the Registration Convention).

the object following default of the debtor/owner. The identity of the responsible party (here, the secured party) could be found by searching the registry of international interests, which would reveal the name of the secured party – information that would not be available through the UN register.

3.5. *The Convention's Relation to the Rescue and Return Agreement*

Most of the discussion in existing literature about the intersection of the Cape Town Convention and existing space law focuses on the Liability Convention and the Registration Convention. However, there are also several scenarios in which the Rescue and Return Agreement (as well as the provisions of the Outer Space Treaty regarding the duty to return errant space objects) may be implicated in a transaction governed by the Cape Town Convention.[129] Most of these scenarios involve the operation of the Rescue and Return Agreement to the benefit of the holder of an international interest, while other situations may arise in which the Rescue and Return Agreement may operate in a manner that would be detrimental to the interests of a creditor.[130]

The following sections provide an overview of the international law requiring the return of errant space objects to the "launching authority," the rescue of spacecraft crew, and other related obligations. The intersections of these obligations with the Cape Town Convention and the Space Assets Protocol will then be explored to determine how the obligations under these space treaties may enhance or interfere with the operation of the Cape Town Convention.

In all of the instances discussed below, it should be kept in mind that the duties under the Rescue and Return Agreement, as well as the Outer Space Treaty, are only triggered with respect to those spacecraft or other space objects that have been "launched into outer space."[131] This means that creditors cannot look to the Rescue and Return Agreement or Outer Space

[129] This intersection between existing space law and the Cape Town Convention has been discussed by Larsen and Heilbock. Larsen and Heilbock, *supra* note 74, at 719.

[130] For a thorough discussion of the duty to rescue astronauts and return space objects see Sundahl, *supra* note 39.

[131] Rescue and Return Agreement, Preamble, First Recital; Art. 5(3); Outer Space Treaty, Art. VIII.

Treaty for assistance in recovering errant space assets unless the assets have been launched. For example, a satellite warehoused in a foreign state prior to launch will have to be recovered by the creditor without the aid of these treaties.

3.5.1. *An Overview of the Duty to Return Space Objects and Rescue Personnel*

The duty to return space objects was originally contained in Paragraph 7 of the 1962 Declaration of Principles and in Article VIII the 1968 Outer Space Treaty.[132] Just one year after the Outer Space Treaty was completed, the Rescue and Return Agreement was concluded in order to elaborate upon the duty to rescue and return that had been established in the Outer Space Treaty.[133] The fact that this treaty was drafted before the drafting of the treaties regarding liability and registration were taken up reflects the importance attached by the United States and the Soviet Union to the duty to rescue astronauts and return lost space objects. Although the duty to rescue astronauts could be said to be purely humanitarian in spirit, the duty to return lost space objects was motivated by the desire to reclaim sensitive technology that had fallen out of the control of the launching state.

Regarding the duty to return errant space objects, Article VIII of the Outer Space Treaty provides as follows:[134]

> [O]bjects or component parts found beyond the limits of the State Party to the Treaty on whose registry they are carried shall be returned to that State Party, which shall, upon request, furnish identifying data prior to their return.

The Outer Space Treaty thus requires the return of any space object to the state of registry whenever the object is found outside of that state's jurisdiction, regardless of the circumstances leading to the discovery of the object. Article 5 of the Rescue and Return Agreement restates this duty to return the space objects of another state, with some modifications:[135]

---

132 Declaration of Principles, para. 7; Outer Space Treaty, Art. VIII.

133 The Moon Agreement also addresses the duty to return space objects by explicitly extending the provisions of the Rescue and Return Agreement to objects found on the Moon or other celestial bodies. Moon Agreement, Art. 12(2).

134 Outer Space Treaty, Art. VIII.

135 Rescue and Return Agreement, Art. 5(3).

> Upon request of the launching authority, objects launched into outer space or their component parts found beyond the territorial limits of the launching authority shall be returned to or held at the disposal of representatives of the launching authority, which shall, upon request, furnish identifying data prior to their return.

This expression of the duty to rescue modifies Article VIII of the Outer Space Treaty in three respects. First, the space object (or component parts) must be returned to the launching authority rather than state on whose registry the object is carried. The term "launching authority" is defined in Article 6 of the Rescue and Return Agreement:[136]

> For the purposes of this Agreement, the term "launching authority" shall refer to the State responsible for launching, or, where an international intergovernmental organization is responsible for launching, that organization, provided that that organization declares its acceptance of the rights and obligations provided for in this Agreement and a majority of the States members of that organization are Contracting Parties to this Agreement and to the Treaty on Principles Governing the Activities of States in the Exploration and Use of Outer Space, including the Moon and Other Celestial Bodies.

In short, an errant space object must be returned to the state (or, under the conditions stated, the international intergovernmental organization) that is "responsible for launching" the object. This rule raises some questions, such as whether a state can be "responsible" for the launching of a space object by a private company and whether any state is "responsible" for the launching of an object from a place outside the territory of any state, such as the high seas. Whether the launching authority will be the state from which the object is launched, the state that commissions the launch, or the state of incorporation of the company that commissions the launch is still open to debate and will be settled as these issues come to court. Second, the Rescue and Return Agreement requires that the launching authority make an official request before the duty to return is triggered. Finally, the Rescue and Return Agreement requires the state holding the space object to either return the object or hold the object "at the disposal of representatives of the launching authority", that is, hold the object for collection by the state that is responsible for launching. These last two modifications are rather inconsequential as the request for return is a simple matter (although it is an added hurdle for private companies to ask their

---

136 Id. Art. 6.

governments to makes such a request) and the option for the finding state to hold the object for collection only shifts to costs of transport to the launching authority.

Article 5 of the Rescue and Return Agreement also expands upon the duty to return errant space objects in three other ways. First, Article 5 requires a state "which receives information or discovers that a space object or its component parts has returned to Earth in territory under its jurisdiction or on the high seas or in any other place not under the jurisdiction of any State" to notify the launching authority and the Secretary-General of the United Nations.[137] Second, Article 5 requires a state on whose territory a spacecraft lands to "take such steps as it finds practicable to recover the object" upon the request of the launching state.[138] Finally, Article 5 places the cost of recovery and return upon the launching authority.[139]

The Outer Space Treaty and the Rescue and Return Agreement also require efforts to rescue the people on board a spacecraft. Article V of the Outer Space Treaty requires states to give astronauts "all possible assistance in the event of accident, distress, or emergency landing on the territory of another State Party or on the high seas."[140] Article 2 of the Rescue and Return Agreement requires that a state "immediately take all possible steps" to rescue spaceflight personnel in the event of an unintended landing in that state's territory.[141] The text of Article 2 is reproduced here:[142]

> If, owing to accident, distress, emergency or unintended landing, the personnel of a spacecraft land in territory under the jurisdiction of a Contracting Party, it shall immediately take all possible steps to rescue them and render them all necessary assistance. It shall inform the launching authority and also the Secretary-General of the United Nations of the steps it is taking and of their progress. If assistance by the launching authority would help to effect a prompt rescue or would contribute substantially to the effectiveness of search and rescue operations, the launching authority shall cooperate with

137 Id. Art. 5(1). Although there is a notification requirement in the Outer Space Treaty, it only requires states to inform other states of "any phenomenon ... which could constitute a danger to the life or health of astronauts." Outer Space Treaty, Art. V.

138 Id. Art. 5(2).

139 Id. Art. 5(5).

140 Id. The language of Article V closely tracks the wording of Paragraph 9 of the *Declaration of Legal Principles Governing the Activities of States in the Exploration and Use of Outer Space*, G.A. Res. 1962 (XVIII), U.N. Doc. A/RES/1962 (Dec. 24, 1963), 3 I.L.M. 157.

141 Rescue and Return Agreement, Art. 2.

142 Id.

> the Contracting Party with a view to the effective conduct of search and rescue operations. Such operations shall be subject to the direction and control of the Contracting Party, which shall act in close and continuing consultation with the launching authority.

Article 3 of the Rescue and Return Agreement complements Article 2 by addressing accidents that occur outside of a state's jurisdiction and provides that if a state discovers that "the personnel of a spacecraft have alighted on the high seas or in any other place not under the jurisdiction of any State, those Contracting Parties which are in a position to do so shall, if necessary, extend assistance in search and rescue operations."[143] It should be noted that the duty to rescue in the circumstances under Article 3 does not require that the landing be due to any emergency. The text of Article 3 of the Rescue and Return Agreement in its entirety is reproduced here:[144]

> If information is received or it is discovered that the personnel of a spacecraft have alighted on the high seas or in any other place not under the jurisdiction of any State, those Contracting Parties which are in a position to do so shall, if necessary, extend assistance in search and rescue operations for such personnel to assure their speedy rescue.

The broad duty imposed on states to engage in search and rescue operations in the event of a downed spacecraft could benefit not only the personnel on board the spacecraft, but will ultimately benefit a company operating the spacecraft (and any creditors of the company). Safety is a critical concern of the flying public and the spaceflight companies. The rescue obligations imposed on all signatory states will contribute to the safety of such company's operations and could help the company maintain solvency in the event of an accident.

3.5.2. *How the Rescue and Return Agreement Can Protect the Interests of a Creditor*

The various duties and obligations placed on states under the Rescue and Return Agreement could benefit a creditor who holds an international interest in a space asset that has landed in a foreign territory. First, the requirement that states notify the launching authority and the UN about the landing of a space object could help a creditor locate a space asset in

143 Id. Art. 3.
144 Id.

which it has an interest. Second, the duty imposed on states to retrieve space objects may assist a creditor in preserving the value of a space asset and gaining possession of the asset. Third, the duty to rescue personnel on board a space object may not only result in the retrieval of a space asset, but may also preserve the safety of any crew or passengers on board – which would have a beneficial effect on the continued operations of the debtor. Finally, the duty of a state to return a space object to the "launching authority" may benefit a creditor if the launching authority has ratified the Cape Town Convention and the asset is therefore returned to a state where the Convention can be applied.

The duty of a state that is a party to the Rescue and Return Agreement to notify the launching authority upon receiving information of the landing of a space object in its territory or on the high seas could benefit a creditor because such notice could assist in the retrieval and preservation of the space asset (as well as any personnel on board). In the event that a spacecraft were lost upon reentry, any information provided by a state party to help save any personnel on board could be critical in maintaining the continued success of the debtor company who relies on its safety record to attract business (from tourists or other human flight participants). The continued success of the debtor's business operations are, of course, always of primary importance to a creditor that is expecting a steady stream of payments from the debtor. A lessor or conditional seller will also be interested in preserving any space asset that would be found (in operational condition) on Earth so that the asset could be retrieved and protected in anticipation of the day when the lessor or seller might regain possession of the asset. Although a reentering satellite would not return to the planet in operation condition, a space plane that was forced down and unable to communicate might only be retrieved due to notification of a state that received information of its descent. In the event that the debtor is in default, a creditor with a security interest could likewise benefit from information about the location of lost collateral. Of course, the state with information of the landing would only be required to notify the launching authority and Secretary General of the UN – any benefit flowing from this notification would only arise if these parties passed notice on to the creditor so that action might then be taken by the creditor or debtor to retrieve the asset.

A creditor with an interest in preserving an asset would likewise benefit from the duty imposed on states party to the Rescue and Return Agreement to retrieve the asset in the event that it lands in their territory. Although satellites that reenter the atmosphere and return to Earth will not survive the landing, a reusable space capsule or a space plane that has been forced down in a foreign country due to an emergency may very well make the landing without suffering damage. The retrieval of the asset would be the first step in preserving the value of the asset, provided that the asset would ultimately come back into the hands of the debtor or creditor. This assistance from a foreign state in the recovery of an errant spacecraft may also take place under the Outer Space Treaty and the Rescue and Return Agreement insofar as these treaties require any states party to these treaties to take all possible actions to rescue the personnel of a spacecraft that has landed regardless of where such landing takes place (with some qualifications).[145] A subsidiary benefit of the duty to rescue is that it may result in the retrieval of the spacecraft as well – to the benefit of any creditor with an interest in the spacecraft. And more importantly, the rescue of crew and passengers in the event of an accident will contribute to the safety record of the spaceflight operation. In addition to the purely humanitarian benefit of such rescues, a healthy safety record will be a critical factor in the survival of any spaceflight company and this should be of interest to any creditor that relies on the continued viability of its debtor.

Another benefit of the Rescue and Return Agreement would flow from the duty of states party to the treaty to return a space object to the launching authority. The benefit of this duty to return errant space objects is not as significant if both the finding state and the launching authority are parties to the Cape Town Convention and the Space Assets Protocol. In this case, even if the finding state did not return the asset to the launching authority, the creditor could proceed to enforce its international interest against the asset in the courts of the finding state under the Cape Town Convention – and should be able to, for example, enforce a security interest by gaining possession and shipping the asset back to where the auction of the asset will be held.[146] Imagine the scenario, however, in which a space

145 The Outer Space Treaty requires states to "render to [astronauts] ... all possible assistance in the event of accident, distress, or emergency landing on the territory of another State Party or on the high seas." Outer Space Treaty, Art. V. The language of Article V of the Outer Space Treaty closely tracks the wording of Paragraph 9 of the Declaration of Legal Principles. Declaration of Principles, para. 9.

146 Some complications may arise in this scenario. For example, the export control laws of the finding state may prevent the creditor from taking possession of the asset or shipping the asset out of the country. In such a case, the provisions of the Rescue and Return Agreement requiring return of the asset may be trumped by export control laws in the

plane takes off from State X, which is a party to the Cape Town Convention, and is forced down in State Y, which is not a party to the Convention, but is a party to the Rescue and Return Agreement. The creditor with an international interest in the space plane would not be able to proceed against the space plane under the Cape Town Convention in State Y. However, State Y would be required under the Rescue and Return Agreement to return the space plane to State X, which would then allow the creditor to proceed under the Cape Town Convention in State X to repossess the space plane and pursue other remedies.

The shifting of all costs of recovery and return required by the Rescue and Return Agreement to the launching authority will also likely benefit creditors since this assurance that costs will be reimbursed by the launching authority should encourage states finding the space object to carry their duty to retrieve and return the object to the launching authority.

3.5.3. *How the Rescue and Return Agreement Could Harm the Interests of a Creditor*

Although the potential benefits of the duty imposed on states to rescue astronauts and return errant space objects seem to outweigh any disadvantages, some scenarios may arise in which a state's obligation under the Rescue and Return Agreement could lead to a result that is contrary to the interests of a creditor. For example, a spacecraft might be returned to a state that is not friendly to the financier and not a party to the Cape Town Convention – and may therefore impede the exercise of the creditor's remedies under the Convention. Consider the following scenario:

> Company delivers customers to privately-owned space stations by way of reusable space capsules. Bank has a security interest in Company's space capsule which is launched from a launch site in State A. State A is not a party to the Cape Town Convention. On its return to Earth, the space capsule guidance systems malfunction and the capsule lands unexpectedly in State B, which is party to both the Rescue and Return Agreement and the Cape Town Convention. The bad press from the emergency landing resulted in cancellations by future customers. Under financial distress from the loss of business, Company defaults on its obligations to Bank. Bank immediately attempts to

---

domestic courts of the finding state (if the legal system of said state gives primacy to the domestic laws over treaty obligations, as is the case under some circumstances in the United States).

> enforce its international interest under the Cape Town Convention by applying for a court order in State B to allow Bank's repossession of the space capsule. At the same time, Company seeks to avoid the repossession of the capsule by requesting that State A make a diplomatic request of State B to return the capsule to State A, as the launching authority, under Rescue and Return Agreement.

How will the conflict between these two competing treaty obligations be resolved? On the one hand, the Cape Town Convention requires the courts of State B to grant Bank possession. On the other hand, the Rescue and Return Agreement requires State B to return the space capsule to State A. Under the Cape Town Convention, primacy is given to the Rescue and Return Agreement and the court may have to dismiss Bank's case requesting repossession. When the capsule is returned to State A, Bank will have no choice but to proceed against the capsule under the domestic laws of State A (which, without the benefit of the Cape Town Convention, may be inimical to its interests).[147] This obstructionist effect of the Rescue and Return Agreement may be sought out by debtors who could choose to launch from a state that is not a party to the Cape Town Convention in an attempt to use the Rescue and Return Agreement to deprive its creditors from availing themselves of the remedies of the Cape Town Convention. It should be kept in mind that a state is obligated to return a space object to the launching authority even if the landing of the object is not the result of emergency or distress.

The foregoing analysis may lead to a different result if the vehicle involved in the above example were a suborbital spaceplane that is only temporarily in space, and which would therefore be subject to the Aircraft Protocol rather than the Space Assets Protocol.[148] Since the Aircraft Protocol does not give supremacy to the UN space treaties, any provisions in the Convention or Aircraft Protocol that conflicted with the provisions of the Rescue and Return Agreement would likely be resolved in favor of the Convention and Aircraft Protocol under the doctrines of *lex posterior* or *lex specialis*. If this were the case, then Bank in the above example would likely succeed in its motion to take possession of the spaceplane since the

[147] Of course, if Country A were a party to the Cape Town Convention and Protocol, once the asset were returned to State A under the Rescue and Return Agreement, the courts of Country A would have to enforce the creditor's right to enforce its remedies under the Convention. See Larsen, *supra* note 41, at 192.

[148] See *supra* Section 2.3.2.

obligation of State B to grant Bank possession would take priority over its duty to return the spaceplane under the Rescue and Return Agreement.

3.6. *Jurisdiction under the Outer Space Treaty and Registration Convention*

As explained above, jurisdiction is a critical issue for the successful operation of the Cape Town Convention, and the Convention and Protocol clearly address which courts will have jurisdiction to adjudicate issues arising under the Convention.[149] However, Article VIII of the Outer Space Treaty threatens to complicate the jurisdictional rules of the Convention through its explicit grant of jurisdiction over a space object to the state of registry:[150]

> A State Party to the Treaty on whose registry an object launched into outer space is carried shall retain jurisdiction and control over such object, and over any personnel thereof, while in outer space or on a celestial body.

Pursuant to the Registration Convention, when a space object is launched into space, the launching state is required to record the launch in its national registry (as well as provide information about the object to the Secretary-General of the United Nations to be included in the international register).[151] It is this state of registry that then has "jurisdiction and control" over the object.

Given the primacy of the Outer Space Treaty, the question arises as to whether Article VIII of the Outer Space Treaty affects the operation of the jurisdictional provisions of the Cape Town Convention. A separate question also presents itself in connection with Outer Space Treaty's grant of jurisdiction to the state of registry – although this question is one that has existed prior to the conclusion of the Cape Town Convention and only promises to be brought into higher relief as the number of commercial transactions increases under the Convention. This question is whether a change must be made to the immutable grant of "jurisdiction and control"

---

149 See *supra* Section 2.12.5.

150 Outer Space Treaty, Art. VIII.

151 Registration Convention, Art. II(1). If there is more than one "launching state", those states are to decide among themselves which state will be the state of registry (of which there can be only one). Id. Art. II(2).

over a space object to the state of registry when subsequent commercial transfers of the object have deprived the state of registry entirely of any actual control over the object. Both of these questions are addressed in the following sections.

3.6.1. *A Summary of the Convention's Provisions on Jurisdiction*

Without the ability of the appropriate courts to exercise jurisdiction in order to enforce a party's rights under the Cape Town Convention, the Convention would be of no value. Jurisdictional questions arise in a number of situations under transactions covered by the Cape Town Convention and the issue is explicitly addressed in the Convention and the Protocol. However, the Outer Space Treaty, which has primacy over the Convention, also addresses the issue of jurisdiction. This section summarizes the jurisdictional provisions of the Convention and Protocol and then examines how the application of these provisions is affected by the Outer Space Treaty.

Jurisdiction arises in four contexts in the Cape Town Convention. In each case, the type of jurisdiction addressed by the Convention is adjudicative jurisdiction. That is, the Convention establishes which state's courts will have the right to rule on a matter arising under the Convention. First, the question of jurisdiction arises when a party seeks the assistance of a court to enforce a remedy or bring any other claim in court under the Convention. For example, if a secured party seeks a court order to obtain control over a debtor's satellite in the event of debtor's default, the court will only be able to issue and enforce this order if it has jurisdiction over the debtor (*in personam* jurisdiction) or over the asset (*in rem* jurisdiction). The general rule under Article 42 of the Convention is that parties to a transaction have the freedom to select the jurisdiction where a dispute arising under the Convention can be adjudicated.[152] Pursuant to Article 42, any such choice of forum agreement between the parties will confer exclusive jurisdiction on the named forum. Of course, a valid forum selection clause will not guarantee the enforcement of that court's order upon adjudication of the claim. For example, if the assets in question (or the means of controlling the asset) are not located in the state of chosen jurisdiction and the party required to perform is not located in that state (and the state has no actual power of enforcement over the party or the party's assets),

[152] Regarding the ability of parties to choose jurisdiction see *supra* Section 2.12.5.

then the enforcement of the order could be difficult in the event that the party refuses to comply with the order. The court issuing the order would have to rely on the enforcement of the order by a court with powers of enforcement over the delinquent party – a feat that is not always easily accomplished.

Second, prior to the final enforcement of a remedy or adjudication of a claim, a party may seek interim relief under Article 13 of the Convention in those states that have jurisdiction pursuant to the agreement of the parties as well as in certain other states depending on the type of relief being sought.[153] Under Article 43, a state in which the asset is "situated" has jurisdiction to grant *in rem* relief, while a state where the debtor is "situated" has jurisdiction to grant *in personam* relief.

Third, in the event that a debtor enters insolvency, the Protocol contemplates that the state where "the centre of the debtor's main interests is situated" will have jurisdiction over the insolvency proceedings and will compel the debtor to deliver the collateral to the creditor in accordance with the insolvency remedies selected by the state pursuant to Article XXI of the Protocol.[154]

Fourth, under Article XXII of the Protocol other states (provided they have made a declaration to this effect) will be required to exercise their jurisdiction over the debtor or collateral to assist the state of primary insolvency jurisdiction in the enforcement of the creditor's remedies.[155]

3.6.2. *How the Outer Space Treaty Affects the Jurisdictional Provisions of the Convention*

The question for parties engaged in a transaction under the Cape Town Convention, for lawyers who represent such parties, and for courts that are asked to adjudicate issues arising under the Convention is whether Article VIII of the Outer Space Treaty affects the jurisdictional provisions of the Convention and the Protocol. The answer to this question depends on how "jurisdiction" is interpreted in the Outer Space Treaty – a question that has been the subject of considerable academic debate.

---

153 See *supra* Section 2.11.9 regarding courts that have jurisdiction to grant interim relief.

154 See *supra* Section 2.11.10 regarding jurisdiction in the context of insolvency.

155 Id.

As a preliminary matter, the nature of the various forms of jurisdiction must be considered. Jurisdiction is generally described as taking three basic forms of (1) prescriptive jurisdiction, (2) enforcement jurisdiction, and (3) adjudicative jurisdiction.[156] Prescriptive jurisdiction is the right of a state to apply its laws to the "activities, relations, or status of persons, or the interests of persons in things."[157] Enforcement jurisdiction refers to a state's right to enforce (or punish noncompliance with) its laws.[158] Finally, adjudicative jurisdiction is the right of a state's courts to subject persons or things to their adjudicative processes and issue a ruling on a matter.[159] Adjudicative jurisdiction can arise over persons (*in personam* jurisdiction) or over property (*in rem* jurisdiction).[160] In addition, jurisdiction can also be "exclusive" or "concurrent." Exclusive jurisdiction exists when a court has sole jurisdiction over a matter and no other court is able to assert its jurisdiction. In contrast, concurrent jurisdiction describes the situation where multiple courts are able to assert their jurisdiction over the same matter.

One interpretation of Article VIII is that only the state of registry has jurisdiction of any kind over a space object and any judicial disputes regarding that space object. This would mean, in other words, that the state of registry has exclusive jurisdiction to regulate the space object and exclusive jurisdiction to hear a case involving the object.[161] If this approach is

---

156 Ryngaert, C., *Jurisdiction in International Law* 9 (2008). See also, generally, Reisman, W.M. (ed.), *Jurisdiction in International Law* (1999).

157 Ryngaert, *supra* note 156, at 9 (citing §401(a) Restatement (Third) of U.S. Foreign Relations Law). Under customary international law, a state has prescriptive jurisdiction if one of the bases of jurisdiction is present. The bases of prescriptive jurisdiction are (1) territoriality, (2) nationality, (3) effects, (4) national security, (5) passive personality, and (6) universality. Raustiala, K., "The Geography of Justice", 73 Fordham Law Review 2501, 2512 (2005). While territoriality and nationality are widely accepted as bases of prescriptive jurisdiction in domestic legal systems, the other bases are accepted only to varying degrees.

158 Ryngaert, *supra* note 156, at 9.

159 Id. 10.

160 von Mehren, A.T., *Adjudicatory Jurisdiction: General Theories Compared and Evaluated,* 63 Boston Univ. Law Review 279, 286 (1983).

161 See, e.g., Aoki, S., "In Search of the Current Legal Status of the Registration of Space Objects", in *2010 Proceedings of the International Institute of Space Law* at 245, 248 (2011) (stating that "[j]urisdiction arising from the registration shall be comprehensive, and a State of registry is supposed to hold legislative, judicial and, above all, enforcement jurisdiction."). Aoki admits that state practice has not always observed such a far-reaching interpretation of the Article VIII of the Outer Space Treaty, but does not think that such

adopted, then Article VIII would trump all jurisdictional provisions of the Cape Town Convention and Protocol and would permit only the state of registry to issue orders regarding the enforcement of remedies and other matters arising under the Convention. This would be an absurd result that would render the Cape Town Convention a nullity since the successful operation of the Convention requires that a creditor be able to take action in any state that exercises actual control over the object or those parties that control the object. Consider, for example, the following scenario:

> A company in State A has a satellite launched from State B and State B registers the satellite. The satellite is controlled from a ground station in State A (and is therefore "situated" in State A for purposes of seeking interim relief). The construction and launch of the satellite was financed by a bank in State C. The bank believes the company has defaulted and seeks interim relief from a court in State A to ensure that the satellite is not moved to a different orbit.

In this scenario, will the court in State A dismiss the case on grounds that under Article VIII of the Outer Space Treaty only courts in State B have jurisdiction? If so, then interim relief will not be possible except in State B – which has no power over the ground station in State A (unless the court's judgment is subsequently enforced by a court in State A). In this scenario, the Cape Town Convention would be of little value to the creditor.

There are commentators who argue for a narrower interpretation of Article VIII of the Outer Space Treaty. The concept of "jurisdiction" in Article VIII can be more narrowly read in at least two respects. First, even if Article VIII is read as granting jurisdiction of every form to the state of registry, it can be interpreted as a non-exclusive grant of jurisdiction – thus enabling other states to assert jurisdiction when appropriate under other sources of domestic and international law.[162] These sources would include

---

state practice affects the interpretation of the treaty. See also Chatzipanagiotis, M., *The Legal Status of Space Tourists in the Framework of Commercial Suborbital Flights* 48 (2011); Lee, *supra* note 110, at 150–151 (stating that "only one state can exercise jurisdiction over ... satellites."); Larsen, P.B., "The Draft Space Protocol and Jurisdiction over Commercial Space Assets", in *2011 Proceedings of the International Institute of Space Law* 485, 488, 490–91 (2012) (citing Aoki's concept of comprehensive jurisdiction as well as other commentators who argue that Article VIII provides an exclusive grant of jurisdiction to the state of registry). At the end, Larsen concludes that there is uncertainty as to whether Article VIII's grant of jurisdiction is exclusive. Id. at 499.

162 Chatzipanagiotis, *supra* note 110, at 50 (stating that "[i]t has been accepted that the State of registry does not have exclusive jurisdiction."). Bin Cheng's writings on jurisdiction

both customary international law (which permits a state to exercise prescriptive jurisdiction if any of the five bases of prescriptive jurisdiction exists) and treaty law (which would include a grant of jurisdiction under the Cape Town Convention and Protocol). This interpretation of Article VIII is reasonable pursuant to a plain reading of its language, which does not state that the jurisdiction of the launching state is exclusive. Under this narrower reading of Article VIII, although State B in the preceding scenario might have jurisdiction to grant interim relief as the state of registry, State A would also have jurisdiction to grant interim relief under the Cape Town Convention since the satellite is "situated in" State A.

Second, Article VIII can be narrowly interpreted as only granting *prescriptive* jurisdiction to the state of registry over a space object (as well as any activity taking place on board the object).[163] This would make sense in the greater context of the concept that the state of registry "shall retain jurisdiction and control." This is a directive that the state of registry maintain control over the space object and makes clear that the laws of the state of registry shall apply to the object (and any activity taking place on board the object). The actual enforcement of such regulations by the state of registry would require that the state have adjudicative jurisdiction over those parties that control the object, which would require some connection of the parties to the state (or *in rem* jurisdiction over the object itself). It could be argued that Article VIII does not address adjudicative jurisdiction.[164] This interpretation of Article VIII would mean that there is no conflict between the Outer Space Treaty and the Cape Town Convention since the

---

also support the reading of Article VIII as a grant of non-exclusive jurisdiction. Cheng, B., "The Extra-Terrestrial Application of International Law", 18 Current Legal Problems 132 (1965). Cheng explains that the seventh principle of the Declaration of Principles (the forerunner of Article VIII) merely codified customary international law by making clear that the state of registry would retain any jurisdiction over its spacecraft and the personnel on board that it would otherwise have under customary international law. Id. at 141.

163 Hobe et al., *supra* note 52, at 159 (stating that "[t]he legal consequence of jurisdiction and control is the applicability of the national law of the State of registry for the object launched into outer space."); see also Gerhard, M. and Gungaphul-Brocard, K., "The Impact of National Space Legislation on Space Industry Contracts", in Smith and Baumann, *supra* note 114, at 64 (explaining that the grant of jurisdiction to the state of registry "defines the law applicable to space objects.").

164 See, e.g., Larsen, *supra* note 41, at 191 (asserting that the "Outer Space Treaty is not intended to cover the issue of jurisdiction of national courts.").

Outer Space Treaty addresses prescriptive jurisdiction, while the Cape Town Convention addresses adjudicative jurisdiction. In the absence of any conflict, the Convention's jurisdiction provisions would be unaffected by the existing space law. Moreover, Article VI of the Outer Space Treaty suggests that grant of prescriptive jurisdiction under Article VIII is a non-exclusive grant of jurisdiction since the duty to supervise national activity under Article VI presupposes that a state have prescriptive jurisdiction over the activities of one's nationals (as well as the space objects operated by its nationals).[165]

The practitioner should also keep in mind that Article VIII of the Outer Space Treaty will only affect the provisions of the Cape Town Convention and the Protocol if the space asset in question qualifies as a "space object." As discussed above, the definition of space asset is broader than the term space object. For example, a satellite that has not yet been launched will qualify as a space asset, but not a space object. If a creditor seeks a court order permitting the repossession of a warehoused satellite, the creditor can proceed under the jurisdictional provisions of the Cape Town Convention without concern that such provisions will be trumped by the Outer Space Treaty.

As a final point, when handling a transaction involving a vehicle that is only temporarily in space, and is therefore subject to the Aircraft Protocol, this analysis of jurisdictional provisions may reach a different result. Since the Aircraft Protocol does not give supremacy to the UN space treaties, any provisions in the Convention or Aircraft Protocol that conflicted with Article VIII of the Outer Space Treaty would likely be resolved in favor of the Convention and Aircraft Protocol under the doctrines of *lex posterior* or *lex specialis*.

3.6.3. *The State of Registry's Retention of Jurisdiction after Transfer*

The increase in commercial transactions involving space assets that is contemplated by the Cape Town Convention raises another issue that is distinct from the question of whether existing space law affects the

---

165 Hobe et al., *supra* note 52, at 113 (stating that "a State has jurisdiction over any activity that is carried on from its territory as well as over any activity that is carried on by its nationals."); see also id. at 114 (explaining that state practice in the form of national space legislation reflects an understanding that the applicability of the legislation to space activities does not rely on the relevant object being registered by the state).

jurisdictional provisions of the Convention, namely, that jurisdiction remains with the state of registry even after transfer of the object to another state. This issue has been a subject of concern among commentators and space faring nations.[166] The issue arises when the immutable rule of jurisdiction and control found in Article VIII of the Outer Space Treaty faces the reality of commercial space use where changes in ownership and control of a space asset already take place (and are likely to take place more frequently in the future).[167]

As described above, Article VIII of the Outer Space Treaty assigns "jurisdiction and control" of a space object to the state of registry. The treaty does not provide for any change in this assignment of jurisdiction and control, but contemplates that it will perpetually remain with the original state of registry. While this approach makes sense during an age when space objects were launched by states and the objects were not the subject of subsequent commercial transactions that would result in the transfer of ownership and actual control to a new party, the current realities of commercial space use have revealed the flaws in the immutable rule of Article VIII.[168] Although it is fitting for the launching state to have jurisdiction and control over a satellite that it has put into orbit and is operating, once this satellite has been transferred to another state it no longer makes sense for

[166] See, e.g., Chatzipanagiotis, *supra* note 110, at 230; Hermida, J., "Transfer of Satellites in Orbit: An International Law Approach", in *Proceedings of the Forty-Sixth Colloquium on the Law of Outer Space* 189, 190 (2004).

[167] Examples of the transfer of on-orbit satellite include (1) the sale by Telesat Canada of the Anik CI and Anik CII satellites to the Argentine corporation Paracomsat S.A, (2) the sale of the satellite BSB-1A by the United Kingdom to the Swedish company Nordic Satellite AB, and (3) the taking of title by Lloyds of the Indonesian Palapa B satellite following the payment of insurance proceeds to the insured and the subsequent sale of the satellite to a new operator. Larsen, *supra* note 41, 190; Ribbelink, *supra* note 41, at 43. Regarding the sale of the satellite BSB-1A see Gerhard, M., *supra* note 90, at 573. See also Chatzipanagiotis, *supra* note 110, at 229; Hörl, K.-U., and Gungaphul, K., "Problems Related to 'Change of Ownership' with Respect to Registration – The Industry View", in *Proceedings of the Project 2001 Plus Workshop "Current Issues in the Registration of Space Objects"* 63, 74–75 (2005).

[168] This flaw in the Outer Space Treaty has been pointed out by a number of commentators including, e.g., Lyall and Larsen, *supra* note 3, at 92; Kerrest, A., "Remarks on the Notion of the Launching State", in *Proceedings of the Forty-Second Colloquium on the Law of Outer Space* at 308, 308–309 (2000); Ribbelink, *supra* note 41, at 42 (explaining that "seizure of property rights and /or of control may have an effect on the degree in which the launching states can fulfill its obligations [to maintain jurisdiction and control under the Outer Space Treaty]."); Chatzipanagiotis, *supra* note 110, at 230.

the original state of registry to retain jurisdiction and control.[169] In fact, the transferee state should be granted jurisdiction and be required to exercise control since this would reflect the reality that the new owner has *de facto* control of the satellite. The transferee state should also have the prescriptive jurisdiction to regulate the operation of the acquired satellite and should have the duty to control the satellite in order to ensure its safe operation. This argument also applies in the private context where a private company in State X places a satellite in orbit and subsequently sells the satellite to a private company incorporated in State Y. Although State X must retain jurisdiction and control under Article VIII, it would be more reasonable for the duty to control the satellite (through appropriate regulation) to shift to State Y while at the same time granting State Y jurisdiction to apply its laws and regulations to the satellite.[170] In another illustration of this problem, consider a company in State A that has placed a private space laboratory in orbit and State A registers the object. If the company later sells the orbital laboratory to a company in State B, the criminal and intellectual property laws of State A would still apply to activities in the laboratory.

One solution to the problem of jurisdiction remaining with the state of registry following a transfer is to have the state of registry enter into an agreement with the transferee state granting the transferee state jurisdiction over the satellite.[171] Such agreements are expressly permitted under Article II of the Registration Convention in order to transfer jurisdiction between launching states, but there is nothing preventing a non-launching state from accepting jurisdiction under such an agreement as well.[172]

---

169 Despite the disadvantages that may result from a state of registry maintaining a transferred satellite on its registry, it is not unheard of. When the Canadian company Telesat Canada sold the Anik CI and Anik CII satellites to the Argentine company Paracom S.A., the satellite remained on Canada's registry. Hermida, J., *Legal Basis for a National Space Regulation* 64–65 (2004).

170 It should be remembered that under Article VI of the Outer Space Treaty a state still has the responsibility to authorize and supervise the space activity of its nationals – even if it is not the state of registry. See *supra* Section 3.4.1.

171 Hermida, *supra* note 166, at 190. Hermida recommends, in the event that there are multiple launching states, that all launching states enter into a jurisdiction transfer agreement with the transferee state. See also Chatzipanagiotis, *supra* note 110, at 232.

172 Perhaps the best known example of an agreement addressing jurisdiction and control is the Intergovernmental Agreement on the International Space Station (IGA). Although the IGA reiterates the jurisdictional rules set forth in the Outer Space Treaty in certain

Another way of transferring "jurisdiction and control" to the transferee state is for the original state of registry to deregister the space object by removing it from its domestic registry.[173] Although such "de-registration" is not mentioned in the space treaties, there is nothing in the treaties that prohibits the practice.[174] This de-registration is followed by the registration of the object by the transferee state, which would result in the jurisdiction and control being attached to the new state of registry. An example of this process took place in 1997 when four satellites (AsiaSat-1, AsiaSat-2, APSTAR-I and APSTAR-IA) which had been previously carried on the United Kingdom registry were deregistered and then subsequently registered by China.[175] The launch of both AsiaSat-1 and AsiaSat-2 took place in China, but the launches were procured by Asia Satellite Telecommunications Co. Ltd., a Hong Kong company. Similarly, the Hong Kong company APT Satellite Co., Ltd. procured the launch of APSTAR-I and APSTAR-IA on a Chinese Long March 3 rocket. At the time of these launches, Hong Kong was a U.K. territory and therefore, as a launching state that had procured the launch of the satellites, the U.K. registered the satellites. When Hong Kong reverted to Chinese control in 1997, the U.K. and China agreed to change the registry of the satellites. A significant limitation to changing the state of registry is that only launching states have the right to register a space object.[176] As a result, this method of transferring jurisdiction is only feasible when there are multiple launching states with respect to a space

respects (by, for example, stating that each state will retain jurisdiction and control over the elements that it registers), it also sets forth new jurisdictional rules that are not found in the space treaties (such as the provision granting each state jurisdiction over its nationals). Agreement among the Government of Canada, Governments of Member States of the European Space Agency, the Government of Japan, the Government of the Russian Federation, and the Government of the United States of America concerning cooperation on the Civil International Space Station, Jan. 29, 1998, T.I.A.S. No. 12927, Arts. 5(2), 22(2). Regarding jurisdictional issues and the IGA see also Blount, P.J., "Jurisdiction in Outer Space: Challenges of Private Individuals in Space", 33 J. Space. L. 299, 312–313 (2007).

173 Kerrest, *supra* note 168, at 308–309.

174 Lee, *supra* note 107, at 153.

175 Information regarding these registrations can be found through the UN Online Index of Objects Launched into Outer Space where links are provided to the following UN documents: ST/SG/SER.E/333 (April 3, 1998); ST/SG/SER.E/334 (April 3, 1998); and ST/SG/SER.E/475 (November 16, 2005). See also Lyall and Larsen, *supra* note 3, at 92–93.

176 Registration Convention, Art. II.

object and the object is transferred from one launching state to another. In the case of the APSTAR and AsiaSat satellite, transfer to the Chinese registry was possible since China was also a launching state (being the state where the launch took place).[177]

3.7. *The Convention's Relation to the Moon Agreement*

Although the Moon Agreement has not been broadly ratified, it is nevertheless in force with respect to the activities of those states that are parties to the treaty. To this extent, the provisions of the Moon Agreement should be considered by practitioners and courts when faced with an issue under the Cape Town Convention. Most of the issues raised by the Moon Agreement in connection with the Cape Town Convention arise in the context of other space treaties – and so some of the matters discussed in this section will be familiar. However, some novel issues also arise from the Moon Agreement. Before delving into these issues, a brief overview of the Moon Agreement follows.

The Moon Agreement governs activity that takes place on the Moon and any other celestial body, including the "orbits around or other trajectories to or around" such bodies.[178] The provisions of the Moon Agreement address the following matters:

- The duty of all states to comply with international law;[179]
- The use of celestial bodies for peaceful purposes and limits on militarization;[180]
- The use of celestial bodies for the benefit of all countries and the principle of international cooperation;[181]

---

177 Upon the on-orbit transfer of satellites to New Skies N.A., a company incorporated in the Netherlands, the Netherlands transmitted a Note Verbale to the Secretary General of the United Nations stating that although the Netherlands was not the state of registry, it did assert "jurisdiction and control" under Article VIII of the Outer Space Treaty. U.N. Doc. A/AC.105/806 (August 22, 2003). See also Lyall and Larsen, *supra* note 1, at 92–93.

178 Moon Agreement, Art. 1(1) & (2).

179 Id. Art. 2.

180 Id. Art. 3.

181 Id. Art. 4.

- The requirement that parties inform the Secretary-General and the public regarding any missions to a celestial body and share scientific data;[182]
- The freedom of scientific exploration;[183]
- The avoidance of environmental damage;[184]
- The freedom to pursue activities on the surface and below the surface of a celestial body provided that interference with the activities of other states is avoided;[185]
- The establishment of space stations;[186]
- The safeguarding of human life and provision of assistance in the event of an emergency;[187]
- The use of a celestial body's natural resources;[188]
- A state's retention of ownership, jurisdiction, and control over its equipment;[189]
- The obligation to return the assets belonging to another state;[190]
- The duty to supervise national activity and bear responsibility and liability for damage;[191] and
- The right of other states to inspect the activity of a state to ensure compliance.[192]

As discussed in the following paragraphs, the provisions of the Moon Agreement that implicate the Cape Town Convention are those that concern (1) provision of assistance in the event of an emergency, (2) inspection rights, (3) the obligation to return assets, and (4) jurisdiction.

Although the Outer Space Treaty and the Rescue and Return Agreement require that states give assistance to astronauts when in distress, the Moon Agreement takes this obligation a step further by requiring that a state "offer shelter in their stations, installations, vehicles and other facilities to persons in distress" on a celestial body (or in the orbit or trajectory toward

182 Id. Art. 5.
183 Id. Art. 6.
184 Id. Art. 7.
185 Id. Art. 8.
186 Id. Art. 9.
187 Id. Arts. 10, 12(3) & 13.
188 Id. Art. 11.
189 Id. Art. 12(1).
190 Id. Art. 12(2).
191 Id. Art. 14.
192 Id. Art. 15.

a celestial body).[193] Moreover, in an "emergency involving a threat to human life," states have the right to use any equipment belonging to another state without prior permission.[194] Some commentators have suggested that these rights and obligations could interfere with the ability of creditors to exercise their remedies against equipment located on celestial bodies.[195] However, the limited effect of these emergency provisions on the ability of a party to use the equipment over a long-term basis would likely not be significant. Moreover, the creditor would not be disadvantaged by these provisions if the creditor needed to sell a lunar station upon the default of the debtor/owner. Any lunar station would be subject to these emergency provisions and therefore any inconvenience caused by these provisions to the owner would be factored equally into the price of any other lunar station in the market.

In addition to the emergency use provisions, the Moon Agreement grants all states party to the treaty the right to inspect the equipment and installations of other states to ensure compliance with the Moon Agreement (such as compliance with the obligation to avoid environmental harm under Article 7).[196] Although a state must give notice prior to a visit and take precautions to avoid harmful interference with the operation of the facility, the visit could be seen as a potential threat to the free use of the facility by its owner. For example, intellectual property could be at risk of discovery by the visiting party. However, as is true for the provisions of the Moon Agreement calling for the emergency use of equipment, the threat of a brief visit with appropriate notice under Article 15 is unlikely to cause significant harm to the value of an asset.

Like the Rescue and Return Agreement, the Moon Agreement requires that if a state finds an errant space object, the state must return the object to the "launching authority" (as defined in the Rescue and Return

---

193 Id. Art. 10(2). For a discussion of the rescue provisions of the Outer Space Treaty and the Rescue and Return Agreement see Sundahl, *supra* note 37, at 167–171. Article 10 of the Moon Agreement requires that states adopt "all practical measures to safeguard the life and health of persons on the moon." Moon Agreement, Art. 10(1). Moreover, all persons on celestial bodies were to be accorded the benefits of "astronauts" under Article V of the Outer Space Treaty and of "personnel" under the Rescue and Return Agreement. Id. See also Larsen and Heilbock, *supra* note 71, at 719.

194 Moon Agreement, Art. 12(3).

195 Larsen and Heilbock, *supra* note 71, at 716.

196 Moon Agreement, Art. 15.

Agreement). In more specific terms, the Moon Agreement requires that "[v]ehicles, installations and equipment or their component parts found in places other than their intended location shall be dealt with in accordance with article 5 of" the Rescue and Return Agreement.[197] As described above in Section 3.5, the Rescue and Return Agreement requires that space objects found by a state outside of its territory must be returned to the "launching state."[198] As described above, this obligation to return the found object to the launching state can be a benefit or a detriment to the creditor.[199] If a space asset found on the Moon is returned to the control of a launching state that is a party to the Cape Town Convention (or is subject to other applicable law that is friendly to the creditor), this obligation to return the object will benefit the creditor. However, if control of the asset is given to a launching state that is not a party to the Convention, the creditor would not be able to avail itself of the remedies of the Convention in the courts of that launching state.

Like Article 5 of the Rescue and Return Agreement, Article 13 of the Moon Agreement requires that a state party aware of an unintended landing of a space object on a celestial body give notice of the landing to both the launching state and the Secretary-General of the United Nations.[200] This information can be of value to a creditor in the event that an asset serving as collateral in a secured transaction has gone off course and is lost. The assistance of another state in locating the lost collateral could help the creditor gain possession in the event of the debtor's default.

The importance of adjudicative jurisdiction to the operation of the Cape Town Convention has been discussed above in Section 3.6. The recognition and enforcement of the jurisdictional provisions contained in the Cape Town Convention are central to its ability to provide comfort to the creditor. While the application of the Convention's jurisdictional provisions may be challenged by the Outer Space Convention's grant of jurisdiction to the state of registry, this concept is refined by the Moon Agreement, which states that "States Parties shall retain jurisdiction and control over their personnel, vehicles, equipment, facilities, stations and installations on the

197 Id. Art. 12(2).

198 See *supra* Section 3.5.1 regarding the duty to return errant space objects under the Rescue and Return Agreement and the definition of "launching state".

199 Id.

200 Moon Agreement, Art. 13.

Moon."[201] Instead of jurisdiction being granted to the state of registry, jurisdiction under the Moon Agreement would seem to be granted to the state that owns (or perhaps has jurisdiction of some type) over the object. While there may be some debate about the precise interpretation of this provision of the Moon Agreement, the grant of jurisdiction appears to have greater flexibility than the jurisdictional rule of the Outer Space Treaty. Whereas the grant of jurisdiction to the state of registry is rather inflexible, the Moon Agreement moves away from this formulation. Under the Moon Agreement, jurisdiction is not attached to registration – but is instead attached to another linkage to a state, such as ownership or jurisdiction of some variety (as is referenced by "States Parties shall retain jurisdiction and control over *their* personnel, vehicles, equipment, facilities, stations"). The arguments presented above regarding the relaxation of the grant of jurisdiction to the state of registry would apply in the context of the Moon Agreement as well. In particular, there is no conclusive argument in support of the view that the grant of jurisdiction should be exclusive to a particular state. Likewise, the statement that jurisdiction should remain with a particular state should not be narrowly interpreted as an exclusive grant of jurisdiction of every type, adjudicative and prescriptive. Instead, multiple states are likely to enjoy concurrent jurisdiction, including the states that are accorded jurisdiction under the Cape Town Convention and Protocol.

3.8. *The Convention's Relation to the ITU Instruments*

The successful operation of the multiplicity of telecommunications satellites in orbit requires careful coordination of both the orbits occupied by satellites and the radio frequencies used by these satellites in order to ensure that the signals received and sent by these satellites do not interfere with one another. The regulation of orbital and radio frequency use is overseen by the International Telecommunications Union (a specialized organ of the United Nations) and the legal framework is composed of the ITU Constitution, the ITU Convention, and the Radio Regulations.[202] One of

201 Id. Art. 12(1).

202 Constitution and Convention of the International Telecommunication Union (1992), art I, 1825 U.N. Treaty Ser. 1, 333 (1998) (hereinafter ITU Constitution); Instrument amending the Constitution and Convention of the International Telecommunication Union, Final Acts of the Plenipotentiary Conference (1994), 1996 Austl. Treaty Ser. 10.

the core concepts of this legal regime is the requirement in Article 45 of the ITU Constitution that "stations" (which include communications satellites) avoid "harmful interference" with radio frequencies in use by other parties:[203]

> All stations, whatever their purpose, must be established and operated in such a manner as not to cause harmful interference to the radio services or communications of other Member States or of recognized operating agencies, or of other duly authorized operating agencies which carry on a radio service, and which operate in accordance with the provisions of the Radio Regulations.

The ITU's competency to regulate the allocation of orbits in addition to radio frequencies is made explicit in Article 44 of the ITU Constitution, which characterizes orbits and frequencies as scarce resources that are to be allocated by states in a rational, efficient and economic manner:[204]

> In using frequency bands for radio services, Member States shall bear in mind that radio frequencies and any associated orbits, including the geostationary-satellite orbit, are limited natural resources and that they must be used rationally, efficiently and economically, in conformity with the provisions of the Radio Regulations, so that countries or groups of countries may have equitable access to those orbits and frequencies, taking into account the special needs of the developing countries and the geographical situation of particular countries.

The complexities of the ITU procedures for allotting orbital slots and radio frequencies is beyond the scope of this work and can be found in specialized writings on the subject.[205] For the purposes of this work, a simplified overview of the process will suffice.[206] In short, the ITU leaves to its member states the authority to license the use of orbits and radio frequencies. When an operator is seeking an allocation of an orbital position and radio frequency, the operator submits the parameters of the satellite's proposed

---

203 ITU Constitution, Art. 45(1).

204 ITU Constitution, Art. 44(2). See also Hertzfeld and von der Dunk, *supra* note 107, at 86–87 (explaining that the "use of frequencies is inseparable from the orbital slot, orbits or orbital planes where the satellite intending to use those frequencies will be located."); Lochan, *supra* note 39, at 50.

205 A concise, yet thorough, description of the ITU filing process can be found in Baumann, I. and Dodel, H., "The ITU Filing of Satellite Systems", in Smith, L.J., and Baumann, I. (eds.), *Contracting for Space* at 367, 367–381 (2011).

206 Lyall and Larsen, *supra* note 1, at 230–234; Hertzfeld and von der Dunk, *supra* note 107, at 86–87; Baumann and Dodel, *supra* note 201, at 375–380.

orbital position, radio frequencies, and any other required information to the appropriate state agency. The state agency will examine the proposed parameters to determine whether they are in accordance with the Radio Regulations and whether there is any threat of interference with existing frequencies. The state agency then transmits the parameters of the allocation to the ITU, which publishes these parameters in its Space International Frequency Information Circular.[207] In the event of potential harmful interference with another operator, consultation and negotiation takes place to modify the proposed allocation in order to avoid the interference.

When considering the intersection between the ITU instruments and the Cape Town Convention, the question is whether the regulatory regime created by the ITU instruments and related domestic laws will interfere with the exercise of remedies by a creditor. In this regard, the critical point to understand is that the use of a particular orbital slot and radio frequency utilized by a satellite may be permitted only pursuant to a license issued under the domestic law of the state receiving the signal.[208] In addition, a creditor must understand that the value of a communications satellite is likely to rely to a large extent on the ability of the satellite to occupy a particular orbital slot and use a particular frequency that is permitted under existing licenses. In short, communications satellite will have little value as collateral unless these licenses can be transferred to a new operator following the transfer of the satellite itself (which transfer might take place, for example, upon the sale of a satellite by a secured party upon the default of the debtor).

The issuance and transfer of licenses for the use of an orbital slot or radio frequency is a matter of applicable domestic law.[209] The Protocol makes this clear in Article XXVI, which states that nothing in the Protocol detracts from a state's right to regulate the transfer of licenses of any kind and, more specifically, a state's right to authorize "the use of orbital positions and frequencies in relation to space assets."[210] This fact that the Protocol does not affect existing laws regarding the use of orbits and frequencies explains

207 The Space International Frequency Information Circular can be accessed online at www.itu.int/sns/demowic.html.

208 See Lochan, *supra* note 39, at 6–50.

209 For information regarding the government agencies that have the authority to issue licenses relating to telecommunications in particular states and an overview of the laws and processes that apply in those states see, generally, Jakhu, *supra* note 3.

210 Space Assets Protocol, Arts. XXVI (2)(a) & (b).

why the ITU has stated that the Cape Town Convention does not interfere with or contradict the operation of the ITU rules and procedures.[211]

In light of the foregoing, when a party takes title or control of a satellite through the mechanisms of the Cape Town Convention, the party should be aware of the following ways in which the use of the satellite may be affected directly or indirectly by the ITU instruments:

- The party may have to seek to have one or more licenses regarding the use of orbits or radio frequencies transferred or modified, which transfer or modification may be denied. This will likely be true even if the satellite remains in the same orbit and continues operations without any changes.[212]
- If the new operator is granted a license, the operator must operate in compliance with the terms of the license and avoid harmful interference with other transmissions.
- A state may limit foreign ownership of licenses to use radio frequencies.[213]
- The state that granted the licenses has the right to suspend or terminate transmissions by the satellite under the ITU Constitution under certain circumstances.[214]
- The ITU Constitution establishes certain priorities for use of telecommunications services that may affect the operation of the satellite.[215]

---

211 van Fenema, *supra* note 39, at 275 (explaining that the ITU indicated "that it saw neither overlap nor contradiction between the [Cape Town] Convention and the draft Protocol, on the one hand, and the ITU Constitution, Convention and Radio Regulations, on the other."); Ribbelink, *supra* note 39, 39 & note 3; Larsen, *supra* note 38, at 7–8 (explaining that Space Assets Protocol Working Group, ITU, and UNCOPUOS found no conflict between the Cape Town Convention and the ITU instruments).

212 In the event that a satellite is moved, the new operator should keep in mind that the same radio frequency may cause interference when the satellite is in one slot but not in another slot. Hertzfeld and von der Dunk, *supra* note 107, at 86–87.

213 For example, Chinese law requires that a telecommunications services operator have majority stock ownership by Chinese nationals. Jakhu, *supra* note 3, at 254; see also Lochan, *supra* note 39, at 6–50.

214 Under Article 34 of the ITU Constitution, states have the right "to cut off, in accordance with their national law, any other private telecommunications which may appear dangerous to the security of the State or contrary to its laws, to public order or to decency." ITU Constitution, Art. 34. Article 35 gives a state the "right to suspend the international telecommunication service ..., provided that it immediately notifies such action to each of the other Member States through the Secretary-General. Id. Art. 35.

215 For example, under Article 40 of the ITU Constitution, priority must be given "to all telecommunications concerning safety of life at sea, on land, in the air or in outer space, as

In order to minimize the detrimental effect of these issues on the value of the collateral, a secured party should include provisions in the underlying agreement with the debtor providing for debtor's consent to the transfer of all related licenses. This consent may serve to facilitate the transfer of the license upon default (although the permission of the appropriate administrative agency may still be required). These regulatory limitations on the use of a communications satellite should also be taken into account when the creditor is determining the real value of collateral when initially deciding whether the debtor is sufficiently collateralized.

---

well as to epidemiological telecommunications of exceptional urgency of the World Health Organization." ITU Constitution, Art. 40. See also Larsen, *supra* note 39, at 192.

ANNEXES

Annex One

Convention on International Interests in Mobile Equipment*

THE STATES PARTIES TO THIS CONVENTION

AWARE of the need to acquire and use mobile equipment of high value or particular economic significance and to facilitate the financing of the acquisition and use of such equipment in an efficient manner,

RECOGNISING the advantages of asset-based financing and leasing for this purpose and desiring to facilitate these types of transaction by establishing clear rules to govern them,

MINDFUL of the need to ensure that interests in such equipment are recognised and protected universally,

DESIRING to provide broad and mutual economic benefits for all interested parties,

BELIEVING that such rules must reflect the principles underlying asset-based financing and leasing and promote the autonomy of the parties necessary in these transactions,

CONSCIOUS of the need to establish a legal framework for international interests in such equipment and for that purpose to create an international registration system for their protection,

TAKING INTO CONSIDERATION the objectives and principles enunciated in existing Conventions relating to such equipment,

HAVE AGREED upon the following provisions:

---

* Reproduced with the permission of Unidroit and the International Civil Aviation Organization.

CHAPTER I

Sphere of application and general provisions

Article 1—Definitions

In this Convention, except where the context otherwise requires, the following terms are employed with the meanings set out below:

(a) "agreement" means a security agreement, a title reservation agreement or a leasing agreement;
(b) "assignment" means a contract which, whether by way of security or otherwise, confers on the assignee associated rights with or without a transfer of the related international interest;
(c) "associated rights" means all rights to payment or other performance by a debtor under an agreement which are secured by or associated with the object;
(d) "commencement of the insolvency proceedings" means the time at which the insolvency proceedings are deemed to commence under the applicable insolvency law;
(e) "conditional buyer" means a buyer under a title reservation agreement;
(f) "conditional seller" means a seller under a title reservation agreement;
(g) "contract of sale" means a contract for the sale of an object by a seller to a buyer which is not an agreement as defined in (a) above;
(h) "court" means a court of law or an administrative or arbitral tribunal established by a Contracting State;
(i) "creditor" means a chargee under a security agreement, a conditional seller under a title reservation agreement or a lessor under a leasing agreement;
(j) "debtor" means a chargor under a security agreement, a conditional buyer under a title reservation agreement, a lessee under a leasing agreement or a person whose interest in an object is burdened by a registrable non-consensual right or interest;
(k) "insolvency administrator" means a person authorised to administer the reorganisation or liquidation, including one authorised on an interim basis, and includes a debtor in possession if permitted by the applicable insolvency law;
(l) "insolvency proceedings" means bankruptcy, liquidation or other collective judicial or administrative proceedings, including interim proceedings, in which the assets and affairs of the debtor are subject to

control or supervision by a court for the purposes of reorganisation or liquidation;

(m) "interested persons" means:

(i) the debtor;

(ii) any person who, for the purpose of assuring performance of any of the obligations in favour of the creditor, gives or issues a suretyship or demand guarantee or a standby letter of credit or any other form of credit insurance;

(iii) any other person having rights in or over the object;

(n) "internal transaction" means a transaction of a type listed in Article 2(2)(a) to (c) where the centre of the main interests of all parties to such transaction is situated, and the relevant object located (as specified in the Protocol), in the same Contracting State at the time of the conclusion of the contract and where the interest created by the transaction has been registered in a national registry in that Contracting State which has made a declaration under Article 50(1);

(o) "international interest" means an interest held by a creditor to which Article 2 applies;

(p) "International Registry" means the international registration facilities established for the purposes of this Convention or the Protocol;

(q) "leasing agreement" means an agreement by which one person (the lessor) grants a right to possession or control of an object (with or without an option to purchase) to another person (the lessee) in return for a rental or other payment;

(r) "national interest" means an interest held by a creditor in an object and created by an internal transaction covered by a declaration under Article 50(1);

(s) "non-consensual right or interest" means a right or interest conferred under the law of a Contracting State which has made a declaration under Article 39 to secure the performance of an obligation, including an obligation to a State, State entity or an intergovernmental or private organisation;

(t) "notice of a national interest" means notice registered or to be registered in the International Registry that a national interest has been created;

(u) "object" means an object of a category to which Article 2 applies;

(v) "pre-existing right or interest" means a right or interest of any kind in or over an object created or arising before the effective date of this Convention as defined by Article 60(2)(a);

(w) "proceeds" means money or non-money proceeds of an object arising from the total or partial loss or physical destruction of the object or its total or partial confiscation, condemnation or requisition;
(x) "prospective assignment" means an assignment that is intended to be made in the future, upon the occurrence of a stated event, whether or not the occurrence of the event is certain;
(y) "prospective international interest" means an interest that is intended to be created or provided for in an object as an international interest in the future, upon the occurrence of a stated event (which may include the debtor's acquisition of an interest in the object), whether or not the occurrence of the event is certain;
(z) "prospective sale" means a sale which is intended to be made in the future, upon the occurrence of a stated event, whether or not the occurrence of the event is certain;
(aa) "Protocol" means, in respect of any category of object and associated rights to which this Convention applies, the Protocol in respect of that category of object and associated rights;
(bb) "registered" means registered in the International Registry pursuant to Chapter V;
(cc) "registered interest" means an international interest, a registrable non-consensual right or interest or a national interest specified in a notice of a national interest registered pursuant to Chapter V;
(dd) "registrable non-consensual right or interest" means a non-consensual right or interest registrable pursuant to a declaration deposited under Article 40;
(ee) "Registrar" means, in respect of the Protocol, the person or body designated by that Protocol or appointed under Article 17(2)(b);
(ff) "regulations" means regulations made or approved by the Supervisory Authority pursuant to the Protocol;
(gg) "sale" means a transfer of ownership of an object pursuant to a contract of sale;
(hh) "secured obligation" means an obligation secured by a security interest;
(ii) "security agreement" means an agreement by which a chargor grants or agrees to grant to a chargee an interest (including an ownership interest) in or over an object to secure the performance of any existing or future obligation of the chargor or a third person;
(jj) "security interest" means an interest created by a security agreement;

(kk) "Supervisory Authority" means, in respect of the Protocol, the Supervisory Authority referred to in Article 17(1);

(ll) "title reservation agreement" means an agreement for the sale of an object on terms that ownership does not pass until fulfilment of the condition or conditions stated in the agreement;

(mm) "unregistered interest" means a consensual interest or non-consensual right or interest (other than an interest to which Article 39 applies) which has not been registered, whether or not it is registrable under this Convention; and

(nn) "writing" means a record of information (including information communicated by teletransmission) which is in tangible or other form and is capable of being reproduced in tangible form on a subsequent occasion and which indicates by reasonable means a person's approval of the record.

Article 2—The international interest

1. This Convention provides for the constitution and effects of an international interest in certain categories of mobile equipment and associated rights.
2. For the purposes of this Convention, an international interest in mobile equipment is an interest, constituted under Article 7, in a uniquely identifiable object of a category of such objects listed in paragraph 3 and designated in the Protocol:
   (a) granted by the chargor under a security agreement;
   (b) vested in a person who is the conditional seller under a title reservation agreement; or
   (c) vested in a person who is the lessor under a leasing agreement.

   An interest falling within sub-paragraph (a) does not also fall within sub-paragraph (b) or (c).
3. The categories referred to in the preceding paragraphs are:
   (a) airframes, aircraft engines and helicopters;
   (b) railway rolling stock; and
   (c) space assets.
4. The applicable law determines whether an interest to which paragraph 2 applies falls within subparagraph (a), (b) or (c) of that paragraph.
5. An international interest in an object extends to proceeds of that object.

Article 3—Sphere of application

1. This Convention applies when, at the time of the conclusion of the agreement creating or providing for the international interest, the debtor is situated in a Contracting State.
2. The fact that the creditor is situated in a non-Contracting State does not affect the applicability of this Convention.

Article 4—Where debtor is situated

1. For the purposes of Article 3(1), the debtor is situated in any Contracting State:
   (a) under the law of which it is incorporated or formed;
   (b) where it has its registered office or statutory seat;
   (c) where it has its centre of administration; or
   (d) where it has its place of business.
2. A reference in sub-paragraph (d) of the preceding paragraph to the debtor's place of business shall, if it has more than one place of business, mean its principal place of business or, if it has no place of business, its habitual residence.

Article 5—Interpretation and applicable law

1. In the interpretation of this Convention, regard is to be had to its purposes as set forth in the preamble, to its international character and to the need to promote uniformity and predictability in its application.
2. Questions concerning matters governed by this Convention which are not expressly settled in it are to be settled in conformity with the general principles on which it is based or, in the absence of such principles, in conformity with the applicable law.
3. References to the applicable law are to the domestic rules of the law applicable by virtue of the rules of private international law of the forum State.
4. Where a State comprises several territorial units, each of which has its own rules of law in respect of the matter to be decided, and where there is no indication of the relevant territorial unit, the law of that State decides which is the territorial unit whose rules shall govern. In the absence of any such rule, the law of the territorial unit with which the case is most closely connected shall apply.

Article 6—Relationship between the Convention and the Protocol

1. This Convention and the Protocol shall be read and interpreted together as a single instrument.
2. To the extent of any inconsistency between this Convention and the Protocol, the Protocol shall prevail.

CHAPTER II

Constitution of an international interest

Article 7—Formal requirements

An interest is constituted as an international interest under this Convention where the agreement creating or providing for the interest:

(a) is in writing;

(b) relates to an object of which the chargor, conditional seller or lessor has power to dispose;

(c) enables the object to be identified in conformity with the Protocol; and

(d) in the case of a security agreement, enables the secured obligations to be determined, but without the need to state a sum or maximum sum secured.

CHAPTER III

Default remedies

Article 8—Remedies of chargee

1. In the event of default as provided in Article 11, the chargee may, to the extent that the charger has at any time so agreed and subject to any declaration that may be made by a Contracting State under Article 54, exercise any one or more of the following remedies:
   (a) take possession or control of any object charged to it;
   (b) sell or grant a lease of any such object;
   (c) collect or receive any income or profits arising from the management or use of any such object.
2. The chargee may alternatively apply for a court order authorising or directing any of the acts referred to in the preceding paragraph.

3. Any remedy set out in sub-paragraph (a), (b) or (c) of paragraph 1 or by Article 13 shall be exercised in a commercially reasonable manner. A remedy shall be deemed to be exercised in a commercially reasonable manner where it is exercised in conformity with a provision of the security agreement except where such a provision is manifestly unreasonable.
4. A chargee proposing to sell or grant a lease of an object under paragraph 1 shall give reasonable prior notice in writing of the proposed sale or lease to:
   (a) interested persons specified in Article 1(m)(i) and (ii); and
   (b) interested persons specified in Article 1(m)(iii) who have given notice of their rights to the chargee within a reasonable time prior to the sale or lease.
5. Any sum collected or received by the chargee as a result of exercise of any of the remedies set out in paragraph 1 or 2 shall be applied towards discharge of the amount of the secured obligations.
6. Where the sums collected or received by the chargee as a result of the exercise of any remedy set out in paragraph 1 or 2 exceed the amount secured by the security interest and any reasonable costs incurred in the exercise of any such remedy, then unless otherwise ordered by the court the chargee shall distribute the surplus among holders of subsequently ranking interests which have been registered or of which the chargee has been given notice, in order of priority, and pay any remaining balance to the chargor.

Article 9—Vesting of object in satisfaction; redemption

1. At any time after default as provided in Article 11, the chargee and all the interested persons may agree that ownership of (or any other interest of the chargor in) any object covered by the security interest shall vest in the chargee in or towards satisfaction of the secured obligations.
2. The court may on the application of the chargee order that ownership of (or any other interest of the chargor in) any object covered by the security interest shall vest in the chargee in or towards satisfaction of the secured obligations.
3. The court shall grant an application under the preceding paragraph only if the amount of the secured obligations to be satisfied by such vesting is commensurate with the value of the object after taking account of any payment to be made by the chargee to any of the interested persons.

4. At any time after default as provided in Article 11 and before sale of the charged object or the making of an order under paragraph 2, the chargor or any interested person may discharge the security interest by paying in full the amount secured, subject to any lease granted by the chargee under Article 8(1)(b) or ordered under Article 8(2). Where, after such default, the payment of the amount secured is made in full by an interested person other than the debtor, that person is subrogated to the rights of the chargee.
5. Ownership or any other interest of the chargor passing on a sale under Article 8(1)(b) or passing under paragraph 1 or 2 of this Article is free from any other interest over which the chargee's security interest has priority under the provisions of Article 29.

Article 10—Remedies of conditional seller or lessor

In the event of default under a title reservation agreement or under a leasing agreement as provided in Article 11, the conditional seller or the lessor, as the case may be, may:

(a) subject to any declaration that may be made by a Contracting State under Article 54, terminate the agreement and take possession or control of any object to which the agreement relates; or
(b) apply for a court order authorising or directing either of these acts.

Article 11—Meaning of default

1. The debtor and the creditor may at any time agree in writing as to the events that constitute a default or otherwise give rise to the rights and remedies specified in Articles 8 to 10 and 13.
2. Where the debtor and the creditor have not so agreed, "default" for the purposes of Articles 8 to 10 and 13 means a default which substantially deprives the creditor of what it is entitled to expect under the agreement.

Article 12—Additional remedies

Any additional remedies permitted by the applicable law, including any remedies agreed upon by the parties, may be exercised to the extent that they are not inconsistent with the mandatory provisions of this Chapter as set out in Article 15.

Article 13—Relief pending final determination

1. Subject to any declaration that it may make under Article 55, a Contracting State shall ensure that a creditor who adduces evidence of default by the debtor may, pending final determination of its claim and to the extent that the debtor has at any time so agreed, obtain from a court speedy relief in the form of such one or more of the following orders as the creditor requests:
   (a) preservation of the object and its value;
   (b) possession, control or custody of the object;
   (c) immobilisation of the object; and
   (d) lease or, except where covered by sub-paragraphs (a) to (c), management of the object and the income therefrom.
2. In making any order under the preceding paragraph, the court may impose such terms as it considers necessary to protect the interested persons in the event that the creditor:
   (a) in implementing any order granting such relief, fails to perform any of its obligations to the debtor under this Convention or the Protocol; or
   (b) fails to establish its claim, wholly or in part, on the final determination of that claim.
3. Before making any order under paragraph 1, the court may require notice of the request to be given to any of the interested persons.
4. Nothing in this Article affects the application of Article 8(3) or limits the availability of forms of interim relief other than those set out in paragraph 1.

Article 14—Procedural requirements

Subject to Article 54(2), any remedy provided by this Chapter shall be exercised in conformity with the procedure prescribed by the law of the place where the remedy is to be exercised.

Article 15—Derogation

In their relations with each other, any two or more of the parties referred to in this Chapter may at any time, by agreement in writing, derogate from or vary the effect of any of the preceding provisions of this Chapter except Articles 8(3) to (6), 9(3) and (4), 13(2) and 14.

CHAPTER IV

The international registration system

Article 16—The International Registry

1. An International Registry shall be established for registrations of:
   (a) international interests, prospective international interests and registrable non-consensual rights and interests;
   (b) assignments and prospective assignments of international interests;
   (c) acquisitions of international interests by legal or contractual subrogations under the applicable law;
   (d) notices of national interests; and
   (e) subordinations of interests referred to in any of the preceding sub-paragraphs.
2. Different international registries may be established for different categories of object and associated rights.
3. For the purposes of this Chapter and Chapter V, the term "registration" includes, where appropriate, an amendment, extension or discharge of a registration.

Article 17—The Supervisory Authority and the Registrar

1. There shall be a Supervisory Authority as provided by the Protocol.
2. The Supervisory Authority shall:
   (a) establish or provide for the establishment of the International Registry;
   (b) except as otherwise provided by the Protocol, appoint and dismiss the Registrar;
   (c) ensure that any rights required for the continued effective operation of the International Registry in the event of a change of Registrar will vest in or be assignable to the new Registrar;
   (d) after consultation with the Contracting States, make or approve and ensure the publication of regulations pursuant to the Protocol dealing with the operation of the International Registry;
   (e) establish administrative procedures through which complaints concerning the operation of the International Registry can be made to the Supervisory Authority;
   (f) supervise the Registrar and the operation of the International Registry;

(g) at the request of the Registrar, provide such guidance to the Registrar as the Supervisory Authority thinks fit;
(h) set and periodically review the structure of fees to be charged for the services and facilities of the International Registry;
(i) do all things necessary to ensure that an efficient notice-based electronic registration system exists to implement the objectives of this Convention and the Protocol; and
(j) report periodically to Contracting States concerning the discharge of its obligations under this Convention and the Protocol.

3. The Supervisory Authority may enter into any agreement requisite for the performance of its functions, including any agreement referred to in Article 27(3).
4. The Supervisory Authority shall own all proprietary rights in the data bases and archives of the International Registry.
5. The Registrar shall ensure the efficient operation of the International Registry and perform the functions assigned to it by this Convention, the Protocol and the regulations.

CHAPTER V

Other matters relating to registration

Article 18—Registration requirements

1. The Protocol and regulations shall specify the requirements, including the criteria for the identification of the object:
   (a) for effecting a registration (which shall include provision for prior electronic transmission of any consent from any person whose consent is required under Article 20);
   (b) for making searches and issuing search certificates, and, subject thereto;
   (c) for ensuring the confidentiality of information and documents of the International Registry other than information and documents relating to a registration.
2. The Registrar shall not be under a duty to enquire whether a consent to registration under Article 20 has in fact been given or is valid.
3. Where an interest registered as a prospective international interest becomes an international interest, no further registration shall be required provided that the registration information is sufficient for a registration of an international interest.

4. The Registrar shall arrange for registrations to be entered into the International Registry data base and made searchable in chronological order of receipt, and the file shall record the date and time of receipt.
5. The Protocol may provide that a Contracting State may designate an entity or entities in its territory as the entry point or entry points through which the information required for registration shall or may be transmitted to the International Registry. A Contracting State making such a designation may specify the requirements, if any, to be satisfied before such information is transmitted to the International Registry.

Article 19—Validity and time of registration

1. A registration shall be valid only if made in conformity with Article 20.
2. A registration, if valid, shall be complete upon entry of the required information into the International Registry data base so as to be searchable.
3. A registration shall be searchable for the purposes of the preceding paragraph at the time when:
   (a) the International Registry has assigned to it a sequentially ordered file number; and
   (b) the registration information, including the file number, is stored in durable form and may be accessed at the International Registry.
4. If an interest first registered as a prospective international interest becomes an international interest, that international interest shall be treated as registered from the time of registration of the prospective international interest provided that the registration was still current immediately before the international interest was constituted as provided by Article 7.
5. The preceding paragraph applies with necessary modifications to the registration of a prospective assignment of an international interest.
6. A registration shall be searchable in the International Registry data base according to the criteria prescribed by the Protocol.

Article 20—Consent to registration

1. An international interest, a prospective international interest or an assignment or prospective assignment of an international interest may be registered, and any such registration amended or extended prior to its expiry, by either party with the consent in writing of the other.

2. The subordination of an international interest to another international interest may be registered by or with the consent in writing at any time of the person whose interest has been subordinated.
3. A registration may be discharged by or with the consent in writing of the party in whose favour it was made.
4. The acquisition of an international interest by legal or contractual subrogation may be registered by the subrogee.
5. A registrable non-consensual right or interest may be registered by the holder thereof.
6. A notice of a national interest may be registered by the holder thereof.

Article 21—Duration of registration

Registration of an international interest remains effective until discharged or until expiry of the period specified in the registration.

Article 22—Searches

1. Any person may, in the manner prescribed by the Protocol and regulations, make or request a search of the International Registry by electronic means concerning interests or prospective international interests registered therein.
2. Upon receipt of a request therefor, the Registrar, in the manner prescribed by the Protocol and regulations, shall issue a registry search certificate by electronic means with respect to any object:
   (a) stating all registered information relating thereto, together with a statement indicating the date and time of registration of such information; or
   (b) stating that there is no information in the International Registry relating thereto.
3. A search certificate issued under the preceding paragraph shall indicate that the creditor named in the registration information has acquired or intends to acquire an international interest in the object but shall not indicate whether what is registered is an international interest or a prospective international interest, even if this is ascertainable from the relevant registration information.

Article 23—List of declarations and declared non-consensual rights or interests

The Registrar shall maintain a list of declarations, withdrawals of declaration and of the categories of nonconsensual right or interest communicated to the Registrar by the Depositary as having been declared by Contracting States in conformity with Articles 39 and 40 and the date of each such declaration or withdrawal of declaration. Such list shall be recorded and searchable in the name of the declaring State and shall be made available as provided in the Protocol and regulations to any person requesting it.

Article 24—Evidentiary value of certificates

A document in the form prescribed by the regulations which purports to be a certificate issued by the International Registry is prima facie proof:

(a) that it has been so issued; and
(b) of the facts recited in it, including the date and time of a registration.

Article 25—Discharge of registration

1. Where the obligations secured by a registered security interest or the obligations giving rise to a registered non-consensual right or interest have been discharged, or where the conditions of transfer of title under a registered title reservation agreement have been fulfilled, the holder of such interest shall, without undue delay, procure the discharge of the registration after written demand by the debtor delivered to or received at its address stated in the registration.
2. Where a prospective international interest or a prospective assignment of an international interest has been registered, the intending creditor or intending assignee shall, without undue delay, procure the discharge of the registration after written demand by the intending debtor or assignor which is delivered to or received at its address stated in the registration before the intending creditor or assignee has given value or incurred a commitment to give value.
3. Where the obligations secured by a national interest specified in a registered notice of a national interest have been discharged, the holder of such interest shall, without undue delay, procure the discharge of the

registration after written demand by the debtor delivered to or received at its address stated in the registration.

4. Where a registration ought not to have been made or is incorrect, the person in whose favour the registration was made shall, without undue delay, procure its discharge or amendment after written demand by the debtor delivered to or received at its address stated in the registration.

Article 26—Access to the international registration facilities

No person shall be denied access to the registration and search facilities of the International Registry on any ground other than its failure to comply with the procedures prescribed by this Chapter.

CHAPTER VI

Privileges and immunities of the Supervisory Authority and the Registrar

Article 27—Legal personality; immunity

1. The Supervisory Authority shall have international legal personality where not already possessing such personality.
2. The Supervisory Authority and its officers and employees shall enjoy such immunity from legal or administrative process as is specified in the Protocol.
3. (a) The Supervisory Authority shall enjoy exemption from taxes and such other privileges as may be provided by agreement with the host State.
   (b) For the purposes of this paragraph, "host State" means the State in which the Supervisory Authority is situated.
4. The assets, documents, data bases and archives of the International Registry shall be inviolable and immune from seizure or other legal or administrative process.
5. For the purposes of any claim against the Registrar under Article 28(1) or Article 44, the claimant shall be entitled to access to such information and documents as are necessary to enable the claimant to pursue its claim.
6. The Supervisory Authority may waive the inviolability and immunity conferred by paragraph 4.

CHAPTER VII

Liability of the Registrar

Article 28—Liability and financial assurances

1. The Registrar shall be liable for compensatory damages for loss suffered by a person directly resulting from an error or omission of the Registrar and its officers and employees or from a malfunction of the international registration system except where the malfunction is caused by an event of an inevitable and irresistible nature, which could not be prevented by using the best practices in current use in the field of electronic registry design and operation, including those related to back-up and systems security and networking.
2. The Registrar shall not be liable under the preceding paragraph for factual inaccuracy of registration information received by the Registrar or transmitted by the Registrar in the form in which it received that information nor for acts or circumstances for which the Registrar and its officers and employees are not responsible and arising prior to receipt of registration information at the International Registry.
3. Compensation under paragraph 1 may be reduced to the extent that the person who suffered the damage caused or contributed to that damage.
4. The Registrar shall procure insurance or a financial guarantee covering the liability referred to in this Article to the extent determined by the Supervisory Authority, in accordance with the Protocol.

CHAPTER VIII

Effects of an international interest as against third parties

Article 29—Priority of competing interests

1. A registered interest has priority over any other interest subsequently registered and over an unregistered interest.
2. The priority of the first-mentioned interest under the preceding paragraph applies:
   (a) even if the first-mentioned interest was acquired or registered with actual knowledge of the other interest; and
   (b) even as regards value given by the holder of the first-mentioned interest with such knowledge.

3. The buyer of an object acquires its interest in it:
   (a) subject to an interest registered at the time of its acquisition of that interest; and
   (b) free from an unregistered interest even if it has actual knowledge of such an interest.
4. The conditional buyer or lessee acquires its interest in or right over that object:
   (a) subject to an interest registered prior to the registration of the international interest held by its conditional seller or lessor; and
   (b) free from an interest not so registered at that time even if it has actual knowledge of that interest.
5. The priority of competing interests or rights under this Article may be varied by agreement between the holders of those interests, but an assignee of a subordinated interest is not bound by an agreement to subordinate that interest unless at the time of the assignment a subordination had been registered relating to that agreement.
6. Any priority given by this Article to an interest in an object extends to proceeds.
7. This Convention:
   (a) does not affect the rights of a person in an item, other than an object, held prior to its installation on an object if under the applicable law those rights continue to exist after the installation; and
   (b) does not prevent the creation of rights in an item, other than an object, which has previously been installed on an object where under the applicable law those rights are created.

Article 30—Effects of insolvency

1. In insolvency proceedings against the debtor an international interest is effective if prior to the commencement of the insolvency proceedings that interest was registered in conformity with this Convention.
2. Nothing in this Article impairs the effectiveness of an international interest in the insolvency proceedings where that interest is effective under the applicable law.
3. Nothing in this Article affects:
   (a) any rules of law applicable in insolvency proceedings relating to the avoidance of a transaction as a preference or a transfer in fraud of creditors; or

(b) any rules of procedure relating to the enforcement of rights to property which is under the control or supervision of the insolvency administrator.

CHAPTER IX

Assignments of associated rights and international interests; rights of subrogation

Article 31—Effects of assignment

1. Except as otherwise agreed by the parties, an assignment of associated rights made in conformity with Article 32 also transfers to the assignee:
   (a) the related international interest; and
   (b) all the interests and priorities of the assignor under this Convention.
2. Nothing in this Convention prevents a partial assignment of the assignor's associated rights. In the case of such a partial assignment the assignor and assignee may agree as to their respective rights concerning the related international interest assigned under the preceding paragraph but not so as adversely to affect the debtor without its consent.
3. Subject to paragraph 4, the applicable law shall determine the defences and rights of set-off available to the debtor against the assignee.
4. The debtor may at any time by agreement in writing waive all or any of the defences and rights of set-off referred to in the preceding paragraph other than defences arising from fraudulent acts on the part of the assignee.
5. In the case of an assignment by way of security, the assigned associated rights revest in the assignor, to the extent that they are still subsisting, when the obligations secured by the assignment have been discharged.

Article 32—Formal requirements of assignment

1. An assignment of associated rights transfers the related international interest only if it:
   (a) is in writing;
   (b) enables the associated rights to be identified under the contract from which they arise; and
   (c) in the case of an assignment by way of security, enables the obligations secured by the assignment to be determined in accordance

with the Protocol but without the need to state a sum or maximum sum secured.

2. An assignment of an international interest created or provided for by a security agreement is not valid unless some or all related associated rights also are assigned.
3. This Convention does not apply to an assignment of associated rights which is not effective to transfer the related international interest.

Article 33—Debtor's duty to assignee

1. To the extent that associated rights and the related international interest have been transferred in accordance with Articles 31 and 32, the debtor in relation to those rights and that interest is bound by the assignment and has a duty to make payment or give other performance to the assignee, if but only if:
   (a) the debtor has been given notice of the assignment in writing by or with the authority of the assignor; and
   (b) the notice identifies the associated rights.
2. Irrespective of any other ground on which payment or performance by the debtor discharges the latter from liability, payment or performance shall be effective for this purpose if made in accordance with the preceding paragraph.
3. Nothing in this Article shall affect the priority of competing assignments.

Article 34—Default remedies in respect of assignment by way of security

In the event of default by the assignor under the assignment of associated rights and the related international interest made by way of security, Articles 8, 9 and 11 to 14 apply in the relations between the assignor and the assignee (and, in relation to associated rights, apply in so far as those provisions are capable of application to intangible property) as if references:

(a) to the secured obligation and the security interest were references to the obligation secured by the assignment of the associated rights and the related international interest and the security interest created by that assignment;
(b) to the chargee or creditor and chargor or debtor were references to the assignee and assignor;

(c) to the holder of the international interest were references to the assignee; and
(d) to the object were references to the assigned associated rights and the related international interest.

Article 35—Priority of competing assignments

1. Where there are competing assignments of associated rights and at least one of the assignments includes the related international interest and is registered, the provisions of Article 29 apply as if the references to a registered interest were references to an assignment of the associated rights and the related registered interest and as if references to a registered or unregistered interest were references to a registered or unregistered assignment.
2. Article 30 applies to an assignment of associated rights as if the references to an international interest were references to an assignment of the associated rights and the related international interest.

Article 36—Assignee's priority with respect to associated rights

1. The assignee of associated rights and the related international interest whose assignment has been registered only has priority under Article 35(1) over another assignee of the associated rights:
   (a) if the contract under which the associated rights arise states that they are secured by or associated with the object; and
   (b) to the extent that the associated rights are related to an object.
2. For the purposes of sub-paragraph (b) of the preceding paragraph, associated rights are related to an object only to the extent that they consist of rights to payment or performance that relate to:
   (a) a sum advanced and utilised for the purchase of the object;
   (b) a sum advanced and utilised for the purchase of another object in which the assignor held another international interest if the assignor transferred that interest to the assignee and the assignment has been registered;
   (c) the price payable for the object;
   (d) the rentals payable in respect of the object; or
   (e) other obligations arising from a transaction referred to in any of the preceding subparagraphs.

3. In all other cases, the priority of the competing assignments of the associated rights shall be determined by the applicable law.

Article 37—Effects of assignor's insolvency

The provisions of Article 30 apply to insolvency proceedings against the assignor as if references to the debtor were references to the assignor.

Article 38—Subrogation

1. Subject to paragraph 2, nothing in this Convention affects the acquisition of associated rights and the related international interest by legal or contractual subrogation under the applicable law.
2. The priority between any interest within the preceding paragraph and a competing interest may be varied by agreement in writing between the holders of the respective interests but an assignee of a subordinated interest is not bound by an agreement to subordinate that interest unless at the time of the assignment a subordination had been registered relating to that agreement.

CHAPTER X

Rights or interests subject to declarations by Contracting States

Article 39—Rights having priority without registration

1. A Contracting State may at any time, in a declaration deposited with the Depositary of the Protocol declare, generally or specifically:
   (a) those categories of non-consensual right or interest (other than a right or interest to which Article 40 applies) which under that State's law have priority over an interest in an object equivalent to that of the holder of a registered international interest and which shall have priority over a registered international interest, whether in or outside insolvency proceedings; and
   (b) that nothing in this Convention shall affect the right of a State or State entity, intergovernmental organisation or other private provider of public services to arrest or detain an object under the laws of that State for payment of amounts owed to such entity, organisation or provider directly relating to those services in respect of that object or another object.

2. A declaration made under the preceding paragraph may be expressed to cover categories that are created after the deposit of that declaration.
3. A non-consensual right or interest has priority over an international interest if and only if the former is of a category covered by a declaration deposited prior to the registration of the international interest.
4. Notwithstanding the preceding paragraph, a Contracting State may, at the time of ratification, acceptance, approval of, or accession to the Protocol, declare that a right or interest of a category covered by a declaration made under sub-paragraph (a) of paragraph 1 shall have priority over an international interest registered prior to the date of such ratification, acceptance, approval or accession.

Article 40—Registrable non-consensual rights or interests

A Contracting State may at any time in a declaration deposited with the Depositary of the Protocol list the categories of non-consensual right or interest which shall be registrable under this Convention as regards any category of object as if the right or interest were an international interest and shall be regulated accordingly. Such a declaration may be modified from time to time.

CHAPTER XI

Application of the Convention to sales

Article 41—Sale and prospective sale

This Convention shall apply to the sale or prospective sale of an object as provided for in the Protocol with any modifications therein.

CHAPTER XII

Jurisdiction

Article 42—Choice of forum

1. Subject to Articles 43 and 44, the courts of a Contracting State chosen by the parties to a transaction have jurisdiction in respect of any claim brought under this Convention, whether or not the chosen forum has a connection with the parties or the transaction. Such jurisdiction shall be exclusive unless otherwise agreed between the parties.

2. Any such agreement shall be in writing or otherwise concluded in accordance with the formal requirements of the law of the chosen forum.

Article 43—Jurisdiction under Article 13

1. The courts of a Contracting State chosen by the parties and the courts of the Contracting State on the territory of which the object is situated have jurisdiction to grant relief under Article 13(1)(a), (b), (c) and Article 13(4) in respect of that object.
2. Jurisdiction to grant relief under Article 13(1)(d) or other interim relief by virtue of Article 13(4) may be exercised either:
   (a) by the courts chosen by the parties; or
   (b) by the courts of a Contracting State on the territory of which the debtor is situated, being relief which, by the terms of the order granting it, is enforceable only in the territory of that Contracting State.
3. A court has jurisdiction under the preceding paragraphs even if the final determination of the claim referred to in Article 13(1) will or may take place in a court of another Contracting State or by arbitration.

Article 44—Jurisdiction to make orders against the Registrar

1. The courts of the place in which the Registrar has its centre of administration shall have exclusive jurisdiction to award damages or make orders against the Registrar.
2. Where a person fails to respond to a demand made under Article 25 and that person has ceased to exist or cannot be found for the purpose of enabling an order to be made against it requiring it to procure discharge of the registration, the courts referred to in the preceding paragraph shall have exclusive jurisdiction, on the application of the debtor or intending debtor, to make an order directed to the Registrar requiring the Registrar to discharge the registration.
3. Where a person fails to comply with an order of a court having jurisdiction under this Convention or, in the case of a national interest, an order of a court of competent jurisdiction requiring that person to procure the amendment or discharge of a registration, the courts referred to in paragraph 1 may direct the Registrar to take such steps as will give effect to that order.
4. Except as otherwise provided by the preceding paragraphs, no court may make orders or give judgments or rulings against or purporting to bind the Registrar.

Article 45—Jurisdiction in respect of insolvency proceedings

The provisions of this Chapter are not applicable to insolvency proceedings.

CHAPTER XIII

Relationship with other Conventions

Article 45 bis*—Relationship with the* United Nations Convention on the Assignment of Receivables in International Trade

This Convention shall prevail over the *United Nations Convention on the Assignment of Receivables in International Trade*, opened for signature in New York on 12 December 2001, as it relates to the assignment of receivables which are associated rights related to international interests in aircraft objects, railway rolling stock and space assets.

Article 46—Relationship with the UNIDROIT Convention on International Financial Leasing

The Protocol may determine the relationship between this Convention and the *UNIDROIT Convention on International Financial Leasing*, signed at Ottawa on 28 May 1988.

CHAPTER XIV

Final provisions

Article 47—Signature, ratification, acceptance, approval or accession

1. This Convention shall be open for signature in Cape Town on 16 November 2001 by States participating in the Diplomatic Conference to Adopt a Mobile Equipment Convention and an Aircraft Protocol held at Cape Town from 29 October to 16 November 2001. After 16 November 2001, the Convention shall be open to all States for signature at the Headquarters of the International Institute for the Unification of Private Law (UNIDROIT) in Rome until it enters into force in accordance with Article 49.
2. This Convention shall be subject to ratification, acceptance or approval by States which have signed it.

3. Any State which does not sign this Convention may accede to it at any time.
4. Ratification, acceptance, approval or accession is effected by the deposit of a formal instrument to that effect with the Depositary.

Article 48—Regional Economic Integration Organisations

1. A Regional Economic Integration Organisation which is constituted by sovereign States and has competence over certain matters governed by this Convention may similarly sign, accept, approve or accede to this Convention. The Regional Economic Integration Organisation shall in that case have the rights and obligations of a Contracting State, to the extent that that Organisation has competence over matters governed by this Convention. Where the number of Contracting States is relevant in this Convention, the Regional Economic Integration Organisation shall not count as a Contracting State in addition to its Member States which are Contracting States.
2. The Regional Economic Integration Organisation shall, at the time of signature, acceptance, approval or accession, make a declaration to the Depositary specifying the matters governed by this Convention in respect of which competence has been transferred to that Organisation by its Member States. The Regional Economic Integration Organisation shall promptly notify the Depositary of any changes to the distribution of competence, including new transfers of competence, specified in the declaration under this paragraph.
3. Any reference to a "Contracting State" or "Contracting States" or "State Party" or "States Parties" in this Convention applies equally to a Regional Economic Integration Organisation where the context so requires.

Article 49—Entry into force

1. This Convention enters into force on the first day of the month following the expiration of three months after the date of the deposit of the third instrument of ratification, acceptance, approval or accession but only as regards a category of objects to which a Protocol applies:
   (a) as from the time of entry into force of that Protocol;
   (b) subject to the terms of that Protocol; and
   (c) as between States Parties to this Convention and that Protocol.

2. For other States this Convention enters into force on the first day of the month following the expiration of three months after the date of the deposit of their instrument of ratification, acceptance, approval or accession but only as regards a category of objects to which a Protocol applies and subject, in relation to such Protocol, to the requirements of sub-paragraphs (a), (b) and (c) of the preceding paragraph.

Article 50—Internal transactions

1. A Contracting State may, at the time of ratification, acceptance, approval of, or accession to the Protocol, declare that this Convention shall not apply to a transaction which is an internal transaction in relation to that State with regard to all types of objects or some of them.
2. Notwithstanding the preceding paragraph, the provisions of Articles 8(4), 9(1), 16, Chapter V, Article 29, and any provisions of this Convention relating to registered interests shall apply to an internal transaction.
3. Where notice of a national interest has been registered in the International Registry, the priority of the holder of that interest under Article 29 shall not be affected by the fact that such interest has become vested in another person by assignment or subrogation under the applicable law.

Article 51—Future Protocols

1. The Depositary may create working groups, in co-operation with such relevant non-governmental organisations as the Depositary considers appropriate, to assess the feasibility of extending the application of this Convention, through one or more Protocols, to objects of any category of high-value mobile equipment, other than a category referred to in Article 2(3), each member of which is uniquely identifiable, and associated rights relating to such objects.
2. The Depositary shall communicate the text of any preliminary draft Protocol relating to a category of objects prepared by such a working group to all States Parties to this Convention, all member States of the Depositary, member States of the United Nations which are not members of the Depositary and the relevant intergovernmental organisations, and shall invite such States and organisations to participate in intergovernmental negotiations for the completion of a draft Protocol on the basis of such a preliminary draft Protocol.

3. The Depositary shall also communicate the text of any preliminary draft Protocol prepared by such a working group to such relevant non-governmental organisations as the Depositary considers appropriate. Such non-governmental organisations shall be invited promptly to submit comments on the text of the preliminary draft Protocol to the Depositary and to participate as observers in the preparation of a draft Protocol.
4. When the competent bodies of the Depositary adjudge such a draft Protocol ripe for adoption, the Depositary shall convene a diplomatic conference for its adoption.
5. Once such a Protocol has been adopted, subject to paragraph 6, this Convention shall apply to the category of objects covered thereby.
6. Article 45 *bis* of this Convention applies to such a Protocol only if specifically provided for in that Protocol.

Article 52—Territorial units

1. If a Contracting State has territorial units in which different systems of law are applicable in relation to the matters dealt with in this Convention, it may, at the time of ratification, acceptance, approval or accession, declare that this Convention is to extend to all its territorial units or only to one or more of them and may modify its declaration by submitting another declaration at any time.
2. Any such declaration shall state expressly the territorial units to which this Convention applies.
3. If a Contracting State has not made any declaration under paragraph 1, this Convention shall apply to all territorial units of that State.
4. Where a Contracting State extends this Convention to one or more of its territorial units, declarations permitted under this Convention may be made in respect of each such territorial unit, and the declarations made in respect of one territorial unit may be different from those made in respect of another territorial unit.
5. If by virtue of a declaration under paragraph 1, this Convention extends to one or more territorial units of a Contracting State:
   (a) the debtor is considered to be situated in a Contracting State only if it is incorporated or formed under a law in force in a territorial unit to which this Convention applies or if it has its registered office or statutory seat, centre of administration, place of business or habitual residence in a territorial unit to which this Convention applies;

(b) any reference to the location of the object in a Contracting State refers to the location of the object in a territorial unit to which this Convention applies; and
(c) any reference to the administrative authorities in that Contracting State shall be construed as referring to the administrative authorities having jurisdiction in a territorial unit to which this Convention applies.

Article 53—Determination of courts

A Contracting State may, at the time of ratification, acceptance, approval of, or accession to the Protocol, declare the relevant "court" or "courts" for the purposes of Article 1 and Chapter XII of this Convention.

Article 54—Declarations regarding remedies

1. A Contracting State may, at the time of ratification, acceptance, approval of, or accession to the Protocol, declare that while the charged object is situated within, or controlled from its territory the chargee shall not grant a lease of the object in that territory.
2. A Contracting State shall, at the time of ratification, acceptance, approval of, or accession to the Protocol, declare whether or not any remedy available to the creditor under any provision of this Convention which is not there expressed to require application to the court may be exercised only with leave of the court.

Article 55—Declarations regarding relief pending final determination

A Contracting State may, at the time of ratification, acceptance, approval of, or accession to the Protocol, declare that it will not apply the provisions of Article 13 or Article 43, or both, wholly or in part. The declaration shall specify under which conditions the relevant Article will be applied, in case it will be applied partly, or otherwise which other forms of interim relief will be applied.

Article 56—Reservations and declarations

1. No reservations may be made to this Convention but declarations authorised by Articles 39, 40, 50, 52, 53, 54, 55, 57, 58 and 60 may be made in accordance with these provisions.

2. Any declaration or subsequent declaration or any withdrawal of a declaration made under this Convention shall be notified in writing to the Depositary.

Article 57—Subsequent declarations

1. A State Party may make a subsequent declaration, other than a declaration authorised under Article 60, at any time after the date on which this Convention has entered into force for it, by notifying the Depositary to that effect.
2. Any such subsequent declaration shall take effect on the first day of the month following the expiration of six months after the date of receipt of the notification by the Depositary. Where a longer period for that declaration to take effect is specified in the notification, it shall take effect upon the expiration of such longer period after receipt of the notification by the Depositary.
3. Notwithstanding the previous paragraphs, this Convention shall continue to apply, as if no such subsequent declarations had been made, in respect of all rights and interests arising prior to the effective date of any such subsequent declaration.

Article 58—Withdrawal of declarations

1. Any State Party having made a declaration under this Convention, other than a declaration authorised under Article 60, may withdraw it at any time by notifying the Depositary. Such withdrawal is to take effect on the first day of the month following the expiration of six months after the date of receipt of the notification by the Depositary.
2. Notwithstanding the previous paragraph, this Convention shall continue to apply, as if no such withdrawal of declaration had been made, in respect of all rights and interests arising prior to the effective date of any such withdrawal.

Article 59—Denunciations

1. Any State Party may denounce this Convention by notification in writing to the Depositary.
2. Any such denunciation shall take effect on the first day of the month following the expiration of twelve months after the date on which notification is received by the Depositary.

3. Notwithstanding the previous paragraphs, this Convention shall continue to apply, as if no such denunciation had been made, in respect of all rights and interests arising prior to the effective date of any such denunciation.

Article 60—Transitional provisions

1. Unless otherwise declared by a Contracting State at any time, the Convention does not apply to a pre-existing right or interest, which retains the priority it enjoyed under the applicable law before the effective date of this Convention.
2. For the purposes of Article 1(v) and of determining priority under this Convention:
   (a) "effective date of this Convention" means in relation to a debtor the time when this Convention enters into force or the time when the State in which the debtor is situated becomes a Contracting State, whichever is the later; and
   (b) the debtor is situated in a State where it has its centre of administration or, if it has no centre of administration, its place of business or, if it has more than one place of business, its principal place of business or, if it has no place of business, its habitual residence.
3. A Contracting State may in its declaration under paragraph 1 specify a date, not earlier than three years after the date on which the declaration becomes effective, when this Convention and the Protocol will become applicable, for the purpose of determining priority, including the protection of any existing priority, to pre-existing rights or interests arising under an agreement made at a time when the debtor was situated in a State referred to in sub-paragraph (b) of the preceding paragraph but only to the extent and in the manner specified in its declaration.

Article 61—Review Conferences, amendments and related matters

1. The Depositary shall prepare reports yearly or at such other time as the circumstances may require for the States Parties as to the manner in which the international regimen established in this Convention has operated in practice. In preparing such reports, the Depositary shall take into account the reports of the Supervisory Authority concerning the functioning of the international registration system.

2. At the request of not less than twenty-five per cent of the States Parties, Review Conferences of States Parties shall be convened from time to time by the Depositary, in consultation with the Supervisory Authority, to consider:
   (a) the practical operation of this Convention and its effectiveness in facilitating the asset-based financing and leasing of the objects covered by its terms;
   (b) the judicial interpretation given to, and the application made of the terms of this Convention and the regulations;
   (c) the functioning of the international registration system, the performance of the Registrar and its oversight by the Supervisory Authority, taking into account the reports of the Supervisory Authority; and
   (d) whether any modifications to this Convention or the arrangements relating to the International Registry are desirable.
3. Subject to paragraph 4, any amendment to this Convention shall be approved by at least a two-thirds majority of States Parties participating in the Conference referred to in the preceding paragraph and shall then enter into force in respect of States which have ratified, accepted or approved such amendment when ratified, accepted, or approved by three States in accordance with the provisions of Article 49 relating to its entry into force.
4. Where the proposed amendment to this Convention is intended to apply to more than one category of equipment, such amendment shall also be approved by at least a two-thirds majority of States Parties to each Protocol that are participating in the Conference referred to in paragraph 2.

Article 62—Depositary and its functions

1. Instruments of ratification, acceptance, approval or accession shall be deposited with the International Institute for the Unification of Private Law (UNIDROIT), which is hereby designated the Depositary.
2. The Depositary shall:
   (a) inform all Contracting States of:
      (i) each new signature or deposit of an instrument of ratification, acceptance, approval or accession, together with the date thereof;
      (ii) the date of entry into force of this Convention;

(iii) each declaration made in accordance with this Convention, together with the date thereof;
(iv) the withdrawal or amendment of any declaration, together with the date thereof; and
(v) the notification of any denunciation of this Convention together with the date thereof and the date on which it takes effect;

(b) transmit certified true copies of this Convention to all Contracting States;
(c) provide the Supervisory Authority and the Registrar with a copy of each instrument of ratification, acceptance, approval or accession, together with the date of deposit thereof, of each declaration or withdrawal or amendment of a declaration and of each notification of denunciation, together with the date of notification thereof, so that the information contained therein is easily and fully available; and
(d) perform such other functions customary for depositaries.

IN WITNESS WHEREOF the undersigned Plenipotentiaries, having been duly authorised, have signed this Convention.

DONE at Cape Town, this sixteenth day of November, two thousand and one, in a single original in the English, Arabic, Chinese, French, Russian and Spanish languages, all texts being equally authentic, such authenticity to take effect upon verification by the Joint Secretariat of the Conference under the authority of the President of the Conference within ninety days hereof as to the conformity of the texts with one another.

Annex Two

Protocol to the Convention on International Interests in Mobile Equipment on Matters Specific to Space Assets*

THE STATES PARTIES TO THIS PROTOCOL,

CONSIDERING it desirable to implement the Convention on International Interests in Mobile Equipment (hereinafter referred to as *the Convention*) as it relates to space assets, in the light of the purposes set out in the preamble to the Convention,

CONSCIOUS of the need to adapt the Convention to meet the particular demand for and the utility of space assets and the need to finance their acquisition and use,

TAKING INTO CONSIDERATION the benefits to all States from expanded space-based services and financing which the Convention and this Protocol may yield,

MINDFUL of the principles of space law, including those contained in the international space treaties of the United Nations and the instruments of the International Telecommunication Union,

RECALLING, for the carrying out of the transfers contemplated by this Protocol, the preeminence of State Party rights and obligations under the international treaties of the United Nations by which the States Parties concerned are bound,

RECOGNISING the continuing development of the international commercial space industry and contemplating the expected benefits of a uniform and predictable regimen governing interests in space assets and in related rights and facilitating asset-based financing of the same,

HAVE AGREED upon the following provisions relating to space assets:

---

* Reproduced with the permission of Unidroit.

CHAPTER I – SPHERE OF APPLICATION AND GENERAL PROVISIONS

Article I—Defined terms

1. – In this Protocol, except where the context otherwise requires, terms used in it have the meanings set out in the Convention.
2. – In this Protocol the following terms are employed with the meanings set out below:
   (a) "debtor's rights" means rights to payment or other performance due or to become due to a debtor by any person with respect to a space asset;
   (b) "guarantee contract" means a contract entered into by a person as a guarantor;
   (c) "guarantor" means a person who, for the purpose of assuring performance of any obligations in favour of a creditor secured by a security agreement or under an agreement, gives or issues a suretyship or demand guarantee or standby letter of credit or other form of credit insurance;
   (d) "insolvency-related event" means:
      (i) the commencement of the insolvency proceedings; or
      (ii) the declared intention to suspend or actual suspension of payments by the debtor where the creditor's right to institute insolvency proceedings against the debtor or to exercise remedies under the Convention is prevented or suspended by law or State action;
   (e) "licence" means any permit, authorisation, concession or equivalent instrument that is granted or issued by, or pursuant to the authority of, a national or intergovernmental or other international body or authority, when acting in a regulatory capacity, to manufacture, launch, control, use or operate a space asset, or relating to the use of orbital positions or the transmission, emission or reception of electromagnetic signals to and from a space asset;
   (f) "obligor" means a person from whom payment or other performance of debtor's rights is due or to become due;
   (g) "primary insolvency jurisdiction" means the Contracting State in which the centre of the debtor's main interests is situated, which for this purpose shall be deemed to be the place of the debtor's statutory seat, or, if there is none, the place where the debtor is incorporated or formed, unless proved otherwise;

(h) "rights assignment" means a contract by which the debtor confers on the creditor an interest (including an ownership interest) in or over the whole or part of existing or future debtor's rights to secure the performance of, or in reduction or discharge of, any existing or future obligation of the debtor to the creditor which under the agreement creating or providing for the international interest is secured by or associated with the space asset to which the agreement relates;
(i) "rights reassignment" means:
(i) a contract by which the creditor transfers to the assignee, or an assignee transfers to a subsequent assignee, the whole or part of its rights and interest under a rights assignment; or
(ii) a transfer of debtor's rights under Article XII(4)(a) of this Protocol;
(j) "space" means outer space, including the Moon and other celestial bodies; and
(k) "space asset" means any man-made uniquely identifiable asset in space or designed to be launched into space, and comprising
(i) a spacecraft, such as a satellite, space station, space module, space capsule, space vehicle or reusable launch vehicle, whether or not including a space asset falling within (ii) or (iii) below;
(ii) a payload (whether telecommunications, navigation, observation, scientific or otherwise) in respect of which a separate registration may be effected in accordance with the regulations; or
(iii) a part of a spacecraft or payload such as a transponder, in respect of which a separate registration may be effected in accordance with the regulations, together with all installed, incorporated or attached accessories, parts and equipment and all data, manuals and records relating thereto.

3.– For the purposes of the definition of "internal transaction" in Article 1(n) of the Convention, a space asset, when not on Earth, is deemed located in the Contracting State which registers the space asset, or on the registry of which the space asset is carried, as a space object under one of the following:
(a) the Treaty on Principles Governing the Activities of States in the Exploration and Use of Outer Space, Including the Moon and Other Celestial Bodies, signed at London, Moscow and Washington, D.C. on 27 January 1967;

(b) the Convention on Registration of Objects Launched into Outer Space, signed at New York on 14 January 1975; or
(c) United Nations General Assembly Resolution 1721 (XVI) B of 20 December 1961.

4. – In Article 43(1) of the Convention and Article XXII of this Protocol, references to a Contracting State on the territory of which an object or space asset is situated shall, as regards a space asset when not on Earth, be treated as references to any of the following:
(a) the Contracting State referred to in the preceding paragraph;
(b) a Contracting State which has issued a licence to operate the space asset; or
(c) a Contracting State on the territory of which a mission control centre for the space asset is located.

Article II—Application of the Convention as regards space assets, debtor's rights and aircraft objects

1. – The Convention shall apply in relation to space assets, rights assignments and rights reassignments as provided by the terms of this Protocol.
2. – The Convention and this Protocol shall be known as the Convention on International Interests in Mobile Equipment as applied to space assets.
3. – This Protocol does not apply to objects falling within the definition of "aircraft objects" under the Protocol to the Convention on International Interests in Mobile Equipment on Matters specific to Aircraft Equipment except where such objects are primarily designed for use in space, in which case this Protocol applies even while such objects are not in space.
4. – This Protocol does not apply to an aircraft object merely because it is designed to be temporarily in space.

Article III – Preservation of rights and interests in a space asset

Ownership of or another right or interest in a space asset shall not be affected by:
(a) the docking of the space asset with another space asset in space;
(b) the installation of the space asset on or the removal of the space asset from another space asset; or
(c) the return of the space asset from space.

Article IV—Application of the Convention to sales; salvage

1. – Article XL of this Protocol and the following provisions of the Convention apply as if references to an agreement creating or providing for an international interest were references to a contract of sale and as if references to an international interest, a prospective international interest, the debtor and the creditor were references to a sale, a prospective sale, the seller and the buyer respectively:

Articles 3 and 4;

Article 16(1)(a);

Article 19(4);

Article 20(1) (as regards registration of a contract of sale or a prospective sale);

Article 25(2) (as regards a prospective sale); and

Article 30.

In addition, the general provisions of Article 1, Article 5, Chapters IV to VII, Article 29 (other than Article 29(3) which is replaced by Article XXIII of this Protocol), Chapter X, Chapter XII (other than Article 43), Chapter XIII and Chapter XIV (other than Article 60) of the Convention shall apply to contracts of sale and prospective sales.

2. – The provisions of this Protocol applicable to rights assignments also apply to a transfer to the buyer of a space asset of rights to payment or other performance due or to become due to the seller by any person with respect to the space asset as if references to the debtor and the creditor were references to the seller and the buyer respectively.

3. – Nothing in the Convention or this Protocol affects any legal or contractual rights of an insurer to salvage recognised by the applicable law. "Salvage" means a legal or contractual right or interest in, relating to or derived from a space asset that vests in the insurer upon the payment of a loss relating to the space asset.

Article V—Formalities, effects and registration of contracts of sale

1. – For the purposes of this Protocol, a contract of sale is one which:
   (a) is in writing;
   (b) relates to a space asset of which the seller has power to dispose; and
   (c) enables the space asset to be identified in conformity with this Protocol.

2. – A contract of sale transfers the interest of the seller in the space asset to the buyer according to its terms.
3. – Registration of a contract of sale remains effective indefinitely. Registration of a prospective sale remains effective unless discharged or until expiry of the period, if any, specified in the registration.

Article VI—Representative capacities

A person may, in relation to a space asset, enter into an agreement or a contract of sale, effect a registration as defined by Article 16(3) of the Convention and assert rights and interests under the Convention in an agency, trust or representative capacity.

Article VII—Identification of space assets

1. – For the purposes of Article 7(c) of the Convention and Articles V and IX of this Protocol, a description of a space asset is sufficient to identify the space asset if it contains:
   (a) a description of the space asset by item;
   (b) a description of the space asset by type;
   (c) a statement that the agreement covers all present and future space assets; or
   (d) a statement that the agreement covers all present and future space assets except for specified items or types.
2. – For the purposes of Article 7 of the Convention, an interest in a future space asset identified in accordance with the preceding paragraph shall be constituted as an international interest as soon as the chargor, conditional seller or lessor acquires the power to dispose of the space asset, without the need for any new act of transfer.

Article VIII—Choice of law

1. – This Article applies unless a Contracting State has made a declaration pursuant to Article XLI(2)(a) of this Protocol.
2. – The parties to an agreement, a contract of sale, a rights assignment or rights reassignment or a related guarantee contract or subordination agreement may agree on the law which is to govern their contractual rights and obligations, wholly or in part.

3. – Unless otherwise agreed, the reference in the preceding paragraph to the law chosen by the parties is to the domestic rules of law of the designated State or, where that State comprises several territorial units, to the domestic law of the designated territorial unit.

Article IX—Formal requirements for rights assignment

A transfer of debtor's rights is constituted as a rights assignment where it is in writing and enables:

(a) the debtor's rights the subject of the rights assignment to be identified;
(b) the space asset to which those rights relate to be identified; and
(c) in the case of a rights assignment by way of security, the obligations secured by the agreement to be determined, but without the need to state a sum or maximum sum secured.

Article X—Effects of rights assignment

1. – A rights assignment made in conformity with Article IX of this Protocol transfers to the creditor the debtor's rights the subject of the rights assignment to the extent permitted by the applicable law.
2. – Subject to paragraph 3, the applicable law shall determine the defences and rights of set-off available to the obligor against the creditor.
3. – The obligor may at any time by agreement in writing waive all or any of the defences and rights of set-off referred to in the preceding paragraph other than defences arising from fraudulent acts on the part of the creditor.

Article XI—Assignment of future rights

A provision in a rights assignment by which future debtor's rights are assigned operates to confer on the creditor an interest in the assigned rights when they come into existence without the need for any new act of transfer.

Article XII—Recording of rights assignment or acquisition by subrogation as part of registration of international interest

1. – The holder of an international interest or prospective international interest in a space asset who has acquired an interest in or over debtor's

rights under a rights assignment or by subrogation may, when registering the international interest or prospective international interest or subsequently by amendment to such registration, record the rights assignment or acquisition by subrogation as part of the registration. Such recording may identify the rights so assigned or acquired either specifically or by a statement that the debtor has assigned, or the holder of the international interest or prospective international interest has acquired, all or some of the debtor's rights, without further specification.

2. – Articles 18, 19, 20(1)-(4), 25(1), (2) and (4) and 30 of the Convention apply in relation to a recording made in accordance with the preceding paragraph as if:
   (a) references to an international interest were references to a rights assignment;
   (b) references to registration were references to the recording of the rights assignment; and
   (c) references to the debtor were references to the obligor.

3. – A search certificate issued under Article 22 of the Convention shall include the particulars recorded under paragraph 1.

4. – Where a rights assignment has been recorded as part of the registration of an international interest which is subsequently transferred in accordance with Articles 31 and 32 of the Convention, the transferee of the international interest acquires:
   (a) all the rights of the creditor under the rights assignment; and
   (b) the right to be shown in the record as assignee under the rights assignment.

5. – Discharge of the registration of an international interest also discharges any recording forming part of that registration under paragraph 1.

Article XIII—Priority of recorded rights assignment

1. – Subject to Article 29(6) of the Convention and paragraph 2 of the present Article, a recorded rights assignment has priority over any other transfer of debtor's rights (whether or not a rights assignment) except a rights assignment previously recorded.

2. – Where a rights assignment is recorded in the registration of a prospective international interest it shall be treated as unrecorded unless and

until the prospective international interest becomes an international interest, in which event the rights assignment has priority as from the time it was recorded provided that the registration was still current immediately before the international interest was constituted as provided by Article 7 of the Convention.

Article XIV—Obligor's duty to creditor

1. – To the extent that the debtor's rights have been assigned to the creditor under a rights assignment, the obligor is bound by the rights assignment, and has a duty to make payment or give other performance to the creditor, if and only if:
   (a) the obligor has been given notice of the rights assignment in writing by or with the authority of the debtor; and
   (b) the notice identifies the debtor's rights.
2. – For the purposes of the preceding paragraph, a notice given by the creditor after the debtor defaults in performance of any obligation secured by a rights assignment is given with the authority of the debtor.
3. – Irrespective of any other ground on which payment or performance by the obligor discharges the obligor from liability, payment or performance shall be effective for this purpose if made in accordance with paragraph 1.
4. – Nothing in this Article shall affect the priority of competing rights assignments.

Article XV—Rights reassignment

1. – Articles IX to XIV of this Protocol apply to a rights reassignment by the creditor or a subsequent assignee. Where those Articles so apply, any references made to the creditor or holder are references to the assignee or subsequent assignee.
2. – A rights reassignment relating to an international interest in a space asset may be recorded only as part of the registration of the assignment of the international interest to the person to whom the rights reassignment was made.

Article XVI—Derogation

The parties may, by agreement in writing, exclude the application of Article XXI of this Protocol and, in their relations with each other, derogate from or vary the effect of any of the provisions of this Protocol except Article XVII(2).

CHAPTER II – DEFAULT REMEDIES, PRIORITIES AND ASSIGNMENTS

Article XVII—Modification of default remedies provisions as regards space assets

1. – Article 8(3) of the Convention shall not apply to space assets. Any remedy given by the Convention in relation to a space asset shall be exercised in a commercially reasonable manner. A remedy shall be deemed to be exercised in a commercially reasonable manner where it is exercised in conformity with a provision of the agreement except where such a provision is manifestly unreasonable.
2. – A chargee giving fourteen or more calendar days' prior written notice of a proposed sale or lease to interested persons shall be deemed to satisfy the requirement of providing "reasonable prior notice" specified in Article 8(4) of the Convention. The foregoing shall not prevent a chargee and a chargor or a guarantor from agreeing to a longer period of prior notice.
3. – Unless otherwise agreed, a creditor may not enforce an international interest in a space asset that is physically linked with another space asset so as to impair or interfere with the operation of the other space asset if an international interest or sale has been registered with respect to the other space asset prior to the registration of the international interest being enforced. For the purposes of this paragraph, a sale or an interest equivalent to an international interest made or arising before the effective date of the Convention, as defined in Article XL of this Protocol, which is registered within three years from that date is deemed to be an international interest or a sale registered at the time of the constitution of the international interest or the sale, as the case may be.

Article XVIII—Default remedies as regards rights assignments and rights reassignments

1. – In the event of default by the debtor under a rights assignment by way of security, Articles 8, 9 and 11 to 14 of the Convention apply in the relations between the debtor and the creditor (and in relation to debtor's rights apply in so far as those provisions are capable of application to intangible property) as if references:
   (a) to the secured obligations and the security interest were references to the obligations secured by the rights assignment and the security interest created by that assignment;
   (b) to the object were references to the debtor's rights.
2. – In the event of default by the assignor under a rights reassignment by way of security, the preceding paragraph applies as if references to the assignment were references to the reassignment.

Article XIX—Placement of data and materials

Subject to Article XXVI of this Protocol, the parties to an agreement may specifically agree for the placement of command codes and related data and materials with another person in order to afford the creditor an opportunity to take possession of, establish control over or operate the space asset.

Article XX—Modification of provisions regarding relief pending final determination

1. – This Article applies only where a Contracting State has made a declaration to that effect under Article XLI(3) of this Protocol and to the extent stated in such declaration.
2. – For the purposes of Article 13(1) of the Convention, "speedy" in the context of obtaining relief means within such number of calendar days from the date of filing of the application for relief as is specified in a declaration made by the Contracting State in which the application is made.
3. – Article 13(1) of the Convention applies with the following being added immediately after sub-paragraph (d):
   "and (e) if at any time the debtor and the creditor specifically agree, sale and application of proceeds therefrom",

and Article 43(2) of the Convention applies with the substitution of "Article 13" for the words "Article 13(1)(d) or other interim relief by virtue of Article 13(4)".

4. – Ownership or any other interest of the debtor passing on a sale under the preceding paragraph is free from any other interest over which the creditor's international interest has priority under the provisions of Article 29 of the Convention.

5. – The creditor and the debtor or any other interested person may agree in writing to exclude the application of Article 13(2) of the Convention.

Article XXI—Remedies on insolvency

1. – This Article applies only where a Contracting State that is the primary insolvency jurisdiction has made a declaration pursuant to Article XLI(4) of this Protocol.

Alternative A

2. – Upon the occurrence of an insolvency-related event, the insolvency administrator or the debtor, as applicable, shall, subject to paragraph 8 and to Article XXVI(2) of this Protocol, give possession of or control over the space asset to the creditor no later than the earlier of:
   (a) the end of the waiting period; and
   (b) the date on which the creditor would be entitled to possession of or control over the space asset if this Article did not apply.

3. – Upon the occurrence of an insolvency-related event, the insolvency administrator or the debtor, as applicable, shall, subject to paragraph 8 and to Article XXVI(2) of this Protocol, give possession of or control over the debtor's rights covered by a rights assignment to the creditor, no later than the earlier of:
   (a) the end of the waiting period; and
   (b) the date on which the creditor would be entitled to possession of or control over the debtor's rights covered by the rights assignment.

4. – For the purposes of this Article, the "waiting period" shall be the period specified in a declaration of the Contracting State which is the primary insolvency jurisdiction.

5. – References in this Article to the "insolvency administrator" shall be to that person in its official, not its personal, capacity.

6. – Unless and until the creditor is given possession of or control over the space asset under paragraph 2 or the debtor's rights under paragraph 3:
   (a) the insolvency administrator or the debtor, as applicable, shall preserve the space asset and maintain it and its value in accordance with the agreement; and
   (b) the creditor shall be entitled to apply for any other forms of interim relief available under the applicable law.
7. – Sub-paragraph (a) of the preceding paragraph shall not preclude the use of the space asset under arrangements designed to preserve the space asset and maintain it and its value.
8. – The insolvency administrator or the debtor, as applicable, may retain possession of and control over the space asset and the debtor's rights covered by a rights assignment where by the time specified in paragraph 2 or paragraph 3 it has cured all defaults other than a default constituted by the opening of insolvency proceedings and has agreed to perform all future obligations under the agreement. A second waiting period shall not apply in respect of a default in the performance of such future obligations.
9. – No exercise of remedies permitted by the Convention or this Protocol may be prevented or delayed after the date specified in paragraph 2 or paragraph 3.
10. – No obligations of the debtor under the agreement may be modified without the consent of the creditor.
11. – Nothing in the preceding paragraph shall be construed to affect the authority, if any, of the insolvency administrator under the applicable law to terminate the agreement.
12. – No rights or interests, except for non-consensual rights or interests of a category covered by a declaration pursuant to Article 39(1) of the Convention, shall have priority in insolvency proceedings over registered interests. This provision shall not derogate from the provisions of Article XXVI(2) of this Protocol.
13. – The Convention as modified by Article XVII of this Protocol shall apply to the exercise of any remedies under this Article.

Alternative B

2. – Upon the occurrence of an insolvency-related event, the insolvency administrator or the debtor, as applicable, upon the request of the

creditor, shall give notice to the creditor within the time specified in a declaration of a Contracting State pursuant to Article XLI(4) of this Protocol whether it will:

(a) cure all defaults other than a default constituted by the opening of insolvency proceedings and agree to perform all future obligations, under the agreement and related transaction documents; or
(b) give the creditor the opportunity to take possession of or control and operation over the space asset, in accordance with the applicable law.

3. – The applicable law referred to in sub-paragraph (b) of the preceding paragraph may permit the court to require the taking of any additional step or the provision of any additional guarantee.
4. – The creditor shall provide evidence of its claims and proof that its international interest has been registered.
5. – If the insolvency administrator or the debtor, as applicable, does not give notice in conformity with paragraph 2, or when it has declared that it will give the creditor the opportunity to take possession of or control and operation over the space asset but fails to do so, the court may permit the creditor to take possession of or control and operation over the space asset upon such terms as the court may order and may require the taking of any additional step or the provision of any additional guarantee.
6. – The space asset shall not be sold pending a decision by a court regarding the claim and the international interest.

Article XXII—Insolvency assistance

1. – This Article applies only where a Contracting State has made a declaration pursuant to Article XLI(2)(b) of this Protocol.
2. – The courts of a Contracting State: (i) in which the space asset is situated; (ii) from which the space asset may be controlled; (iii) in which the debtor is located; (iv) in which the space asset is registered; (v) which has issued a licence in respect of the space asset; or (vi) otherwise having a close connection with the space asset, shall, in accordance with the law of the Contracting State, co-operate to the maximum extent possible with foreign courts and foreign insolvency administrators in carrying out the provisions of Article XXI of this Protocol.

Article XXIII—Modification of priority provisions

1. – The buyer of a space asset under a registered sale acquires its interest in that asset free from an interest subsequently registered and from an unregistered interest, even if the buyer has actual knowledge of the unregistered interest.
2. – The buyer of a space asset under a registered sale acquires its interest in that asset subject to an interest previously registered.

Article XXIV—Modification of assignment provisions

Article 33(1) of the Convention applies with the following being added immediately after sub-paragraph (b):

"and (c) the debtor has consented in writing, whether or not the consent is given in advance of the assignment or identifies the assignee."

Article XXV—Debtor provisions

1. – In the absence of a default within the meaning of Article 11 of the Convention, the debtor shall be entitled to the quiet possession and use of the space asset in accordance with the agreement as against:
   (a) its creditor and the holder of any interest from which the debtor takes free pursuant to Article 29(4)(b) of the Convention or, in the capacity of buyer, Article XXIII(1) of this Protocol, unless and to the extent that the debtor has otherwise agreed; and
   (b) the holder of any interest to which the debtor's right or interest is subject pursuant to Article 29(4)(a) of the Convention or, in the capacity of buyer, Article XXIII(2) of this Protocol, but only to the extent, if any, that such holder has agreed.
2. – Nothing in the Convention or this Protocol affects the liability of a creditor for any breach of the agreement under the applicable law in so far as that agreement relates to space assets.

Article XXVI—Preservation of powers of Contracting States

1. – This Protocol does not affect the exercise by a Contracting State of its authority to issue licences, approvals, permits or authorisations for the launch or operation of space assets or the provision of any service through the use or with the support of space assets.

2. – This Protocol further does not:
   (a) render transferable or assignable any licences, approvals, permits or authorisations which, in accordance with the laws and regulations of the granting Contracting State or the contractual or administrative provisions under which they are granted, may not be transferred or assigned;
   (b) limit the right of a Contracting State to authorise the use of orbital positions and frequencies in relation to space assets; or
   (c) affect the ability of a Contracting State in accordance with its laws and regulations to prohibit, restrict or attach conditions to the placement of command codes and related data and materials pursuant to Article XIX of this Protocol.

3. – Nothing in this Protocol shall be construed so as to require a Contracting State to recognise or enforce an international interest in a space asset when the recognition or enforcement of such interest would conflict with its laws or regulations concerning:
   (a) the export of controlled goods, technology, data and services; or
   (b) national security.

Article XXVII—Limitations on remedies in respect of public service

1. – Where the debtor or an entity controlled by the debtor and a public services provider enter into a contract that provides for the use of a space asset to provide services that are needed for the provision of a public service in a Contracting State, the parties and the Contracting State may agree that the public services provider or the Contracting State may register a public service notice.

2. – For the purposes of this Article:
   (a) "public service notice" means a notice in the International Registry describing, in accordance with the regulations, the services which under the contract are intended to support the provision of a public service recognised as such under the laws of the relevant Contracting State at the time of registration; and
   (b) "public services provider" means an entity of a Contracting State, another entity situated in that Contracting State and designated by the Contracting State as a provider of a public service or an entity recognised as a provider of a public service under the laws of a Contracting State.

3. – Subject to paragraph 9, a creditor holding an international interest in a space asset that is the subject of a public service notice may not, in the event of default, exercise any of the remedies provided in Chapter III of the Convention or Chapter II of this Protocol that would make the space asset unavailable for the provision of the relevant public service prior to the expiration of the period specified in a declaration by a Contracting State as provided by paragraph 4.
4. – A Contracting State shall at the time of ratification, acceptance, approval of, or accession to this Protocol specify by a declaration under Article XLI(1) a period for the purposes of the preceding paragraph not less than three months nor more than six months from the date of registration by the creditor of a notice in the International Registry that the creditor may exercise any such remedies if the debtor does not cure its default within that period.
5. – Paragraph 3 does not affect the ability of a creditor, if so authorised by the relevant authorities, temporarily to operate or ensure the continued operation of a space asset during the period referred to in that paragraph where the debtor is not able to do so.
6. – The creditor shall promptly notify the debtor and the public services provider of the date of registration of its notice under paragraph 3 and of the date of expiry of the period referred to therein.
7. – During the period referred to in paragraph 3:
   (a) the creditor, the debtor and the public services provider shall co-operate in good faith with a view to finding a commercially reasonable solution permitting the continuation of the public service;
   (b) the regulatory authority of a Contracting State that issued a licence required by the debtor to operate the space asset that is the subject of a public service notice shall, as appropriate, give the public services provider the opportunity to participate in any proceedings in which the debtor may participate in that Contracting State, with a view to the appointment of another operator under a new licence to be issued by that regulatory authority; and
   (c) the creditor is not precluded from initiating proceedings with a view to the replacement of the debtor by another person as operator of the space asset concerned in accordance with the rules of the licensing authorities.
8. – Notwithstanding paragraphs 3 and 7, the creditor is free to exercise any of the remedies provided in Chapter III of the Convention or

Chapter II of this Protocol if, at any time during the period referred to in paragraph 3, the public services provider fails to perform its duties under the contract referred to in paragraph 1.

9. – Unless otherwise agreed, the limitation on the remedies of the creditor provided for in paragraph 3 shall not apply in respect of an international interest registered by a creditor prior to the registration of a public service notice pursuant to paragraph 1, provided that:
   (a) the international interest was created pursuant to an agreement made before the conclusion of the contract with the public services provider referred to in paragraph 1; and
   (b) at the time the international interest was registered in the International Registry, the creditor had no knowledge that such a public services contract had been entered into.

10. – The preceding paragraph does not apply if such public service notice is registered no later than six months after the initial launch of the space asset.

CHAPTER III – REGISTRY PROVISIONS RELATING TO INTERNATIONAL INTERESTS IN SPACE ASSETS

Article XXVIII—The Supervisory Authority

1. – The Supervisory Authority shall be designated at, or pursuant to a resolution of, the diplomatic Conference for the adoption of the draft Protocol to the Convention on International Interests in Mobile Equipment on Matters specific to Space Assets, provided that such Supervisory Authority is able and willing to act in such capacity.

2. – The Supervisory Authority and its officers and employees shall enjoy such immunity from legal and administrative process as is provided under the rules applicable to them as an international entity or otherwise.

3. – The Supervisory Authority shall establish a commission of experts, from among persons nominated by the negotiating States and having the necessary qualifications and experience, and entrust it with the task of assisting the Supervisory Authority in the discharge of its functions.

Article XXIX—First regulations

The first regulations shall be made by the Supervisory Authority so as to take effect on the entry into force of this Protocol.

Article XXX—Identification of space assets for registration purposes

A description of a space asset in accordance with the criteria for identification provided by the regulations is necessary and sufficient to identify the space asset for the purposes of registration in the International Registry.

Article XXXI—Designated entry points

A Contracting State may at any time designate an entity or entities in its territory as the entry point or entry points through which there shall or may be transmitted to the International Registry information required for registration other than registration of a notice of a national interest or a right or interest under Article 40 of the Convention in either case arising under the laws of another State.

Article XXXII—Additional modifications to Registry provisions

1. – Article 16 of the Convention applies with the following being added immediately after paragraph 1:
"1 *bis* The International Registry shall also provide for:
(a) the recording of rights assignments and rights reassignments;
(b) the recording of acquisitions of debtor's rights by subrogation;
(c) the registration of public service notices under Article XXVII(1) of the Protocol to the Convention on International Interests in Mobile Equipment on Matters specific to Space Assets; and
(d) the registration of creditors' notices under Article XXVII(4) of the Protocol to the Convention on International Interests in Mobile Equipment on Matters specific to Space Assets."
2. – For the purposes of Article 19(6) of the Convention, the search criteria for space assets shall be the criteria specified in Article XXX of this Protocol.
3. – For the purposes of Article 25(2) of the Convention, and in the circumstances there described, the holder of a registered prospective international interest or a registered prospective assignment of an

international interest shall take such steps as are within its power to procure the discharge of the registration no later than ten calendar days after the receipt of the demand described in such paragraph.

4. – The fees referred to in Article 17(2)(h) of the Convention shall be determined so as to recover the reasonable costs of establishing, operating and regulating the International Registry and the reasonable costs of the Supervisory Authority associated with the performance of the functions, exercise of the powers and discharge of the duties contemplated by Article 17(2) of the Convention.
5. – The centralised functions of the International Registry shall be operated and administered by the Registrar on a twenty-four hour basis.
6. – The insurance or financial guarantee referred to in Article 28(4) of the Convention shall cover the liability of the Registrar under the Convention to the extent provided by the regulations.
7. – Nothing in the Convention shall preclude the Registrar from procuring insurance or a financial guarantee covering events for which the Registrar is not liable under Article 28 of the Convention.

CHAPTER IV – JURISDICTION

Article XXXIII—Waiver of sovereign immunity

1. – Subject to paragraph 2, a waiver of sovereign immunity from jurisdiction of the courts specified in Article 42 or Article 43 of the Convention or relating to enforcement of rights and interests relating to a space asset under the Convention shall be binding and, if the other conditions to such jurisdiction or enforcement have been satisfied, shall be effective to confer jurisdiction and permit enforcement, as the case may be.
2. – A waiver under the preceding paragraph must be in writing and contain a description of the space asset in accordance with Article VII of this Protocol.

CHAPTER V – RELATIONSHIP WITH OTHER CONVENTIONS

Article XXXIV—Relationship with the UNIDROIT Convention on International Financial Leasing

The Convention as applied to space assets shall supersede the *UNIDROIT Convention on International Financial Leasing* in respect of the subject matter of this Protocol, as between States Parties to both Conventions.

Article XXXV—Relationship with the United Nations outer space treaties and instruments of the International Telecommunication Union

The Convention as applied to space assets shall not affect State Party rights and obligations under the existing United Nations outer space treaties or instruments of the International Telecommunication Union.

CHAPTER VI – FINAL PROVISIONS

Article XXXVI—Signature, ratification, acceptance, approval or accession

1. – This Protocol shall be open for signature in Berlin on 9 March 2012 by States participating in the diplomatic Conference for the adoption of the draft Protocol to the Convention on International Interests in Mobile Equipment on Matters specific to Space Assets held in Berlin from 27 February to 9 March 2012. After 9 March 2012 this Protocol shall be open to all States for signature at Rome until it enters into force in accordance with Article XXXVIII.
2. – This Protocol shall be subject to ratification, acceptance or approval by States which have signed it.
3. – Any State which does not sign this Protocol may accede to it at any time.
4. – Ratification, acceptance, approval or accession is effected by the deposit of a formal instrument to that effect with the Depositary.
5. – A State may not become a Party to this Protocol unless it is or becomes also a Party to the Convention.

Article XXXVII—Regional Economic Integration Organisations

1. – A Regional Economic Integration Organisation which is constituted by sovereign States and has competence over certain matters governed by this Protocol may similarly sign, accept, approve or accede to this Protocol. The Regional Economic Integration Organisation shall in that case have the rights and obligations of a Contracting State, to the extent that that Organisation has competence over matters governed by this Protocol. Where the number of Contracting States is relevant in this Protocol, the Regional Economic Integration Organisation shall not count as a Contracting State in addition to its Member States which are Contracting States.

2. – The Regional Economic Integration Organisation shall, at the time of signature, acceptance, approval or accession, make a declaration to the Depositary specifying the matters governed by this Protocol in respect of which competence has been transferred to that Organisation by its Member States. The Regional Economic Integration Organisation shall promptly notify the Depositary in writing of any changes to the distribution of competence, including new transfers of competence, specified in the declaration under this paragraph.
3. – Any reference to a "Contracting State", "Contracting States", "State Party" or "States Parties" in this Protocol applies equally to a Regional Economic Integration Organisation where the context so requires.

Article XXXVIII—Entry into force

1. – This Protocol enters into force between the States which have deposited instruments referred to in sub-paragraph (a) on the later of:
   (a) the first day of the month following the expiration of three months after the date of the deposit of the tenth instrument of ratification, acceptance, approval or accession; and
   (b) the date of the deposit by the Supervisory Authority with the Depositary of a certificate confirming that the International Registry is fully operational.
2. – For other States this Protocol enters into force on the first day of the month following the later of:
   (a) the expiration of three months after the date of the deposit of their instrument of ratification, acceptance, approval or accession; and
   (b) the date referred to in sub-paragraph (b) of the preceding paragraph.

Article XXXIX—Territorial units

1. – If a Contracting State has two or more territorial units in which different systems of law are applicable in relation to the matters dealt with in this Protocol, it may, at the time of signature, ratification, acceptance, approval or accession, make an initial declaration that this Protocol is to extend to all its territorial units or only to one or more of them and may modify its declaration by submitting another declaration at any time.

2. – Any such declaration shall state expressly the territorial units to which this Protocol applies.
3. – If a Contracting State has not made any declaration under paragraph 1, this Protocol shall apply to all territorial units of that State.
4. – Where a Contracting State extends this Protocol to one or more of its territorial units, declarations permitted under this Protocol may be made in respect of each such territorial unit, and the declarations made in respect of one territorial unit may be different from those made in respect of another territorial unit.
5. – In relation to a Contracting State with two or more territorial units in which different systems of law are applicable in relation to the matters dealt with in this Protocol, any reference to the law in force in a Contracting State or to the law of a Contracting State shall be construed as referring to the law in force in the relevant territorial unit.
6. – If a Contracting State has a federal system where the federal legislative power has competence over matters governed by this Protocol, that Contracting State shall have the same rights and obligations over those matters as those Contracting States which do not have a federal system.

Article XL—Transitional provisions

1. – Article 60 of the Convention shall not apply in relation to space assets.
2. – Subject to the second sentence of Article XVII(3) of this Protocol, the Convention does not apply to a right or interest of any kind in or over a space asset created or arising before the effective date of the Convention, which retains the priority it enjoyed under the applicable law before the effective date of the Convention.
3. – For the purposes of this Protocol:
   (a) "effective date of the Convention" means in relation to a debtor the time when the Convention enters into force or the time when the State in which the debtor is situated at the time the right or interest is created or arises becomes a Contracting State, whichever is the later; and
   (b) the debtor is situated in a State where it has its centre of administration or, if it has no centre of administration, its place of business or, if it has more than one place of business, its principal place of business or, if it has no place of business, its habitual residence.

Article XLI—Declarations relating to certain provisions

1. – A Contracting State shall, at the time of ratification, acceptance, approval of, or accession to this Protocol, make a declaration pursuant to Article XXVII(4) of this Protocol.
2. – A Contracting State may, at the time of ratification, acceptance, approval of, or accession to this Protocol, declare:
   (a) that it will not apply Article VIII;
   (b) that it will apply Article XXII.
3. – A Contracting State may, at the time of ratification, acceptance, approval of, or accession to this Protocol, declare that it will apply Article XX wholly or in part. If it so declares with respect to Article XX(2), it shall specify the time-period required thereby.
4. – A Contracting State may, at the time of ratification, acceptance, approval of, or accession to this Protocol, declare that it will apply the entirety of Alternative A, or the entirety of Alternative B of Article XXI and, if so, shall specify the types of insolvency proceeding, if any, to which it will apply Alternative A and the types of insolvency proceeding, if any, to which it will apply Alternative B. A Contracting State making a declaration pursuant to this paragraph shall specify the time-period required by Article XXI.
5. – The courts of Contracting States shall apply Article XXI in conformity with the declaration made by the Contracting State that is the primary insolvency jurisdiction.

Article XLII—Declarations under the Convention

Declarations made under the Convention, including those made under Articles 39, 40, 53, 54, 55, 57 and 58 of the Convention, shall be deemed to have also been made under this Protocol unless stated otherwise.

Article XLIII—Reservations and declarations

1. – No reservations may be made to this Protocol but declarations authorised by Articles XXXIX, XLI, XLII and XLIV may be made in accordance with these provisions.
2. – Any declaration, subsequent declaration or any withdrawal of a declaration made under this Protocol shall be notified in writing to the Depositary.

Article XLIV—Subsequent declarations

1. – A State Party may make a subsequent declaration at any time after the date on which this Protocol has entered into force for it, by notifying the Depositary to that effect.
2. – Any such subsequent declaration shall take effect on the first day of the month following the expiration of six months after the date of receipt of the notification by the Depositary. Where a longer period for that declaration to take effect is specified in the notification, it shall take effect upon the expiration of such longer period after receipt of the notification by the Depositary.
3. – Notwithstanding the previous paragraphs, this Protocol shall continue to apply, as if no such subsequent declaration had been made, in respect of all rights and interests arising prior to the effective date of any such subsequent declaration.

Article XLV—Withdrawal of declarations

1. – Any State Party having made a declaration under this Protocol may withdraw it at any time by notifying the Depositary. Such withdrawal is to take effect on the first day of the month following the expiration of six months after the date of receipt of the notification by the Depositary.
2. – Notwithstanding the previous paragraph, this Protocol shall continue to apply, as if no such withdrawal of declaration had been made, in respect of all rights and interests arising prior to the effective date of any such withdrawal of declaration.

Article XLVI—Denunciations

1. – Any State Party may denounce this Protocol by notification in writing to the Depositary.
2. – Any such denunciation shall take effect on the first day of the month following the expiration of twelve months after the date of receipt of the notification by the Depositary.
3. – Notwithstanding the previous paragraphs, this Protocol shall continue to apply, as if no such denunciation had been made, in respect of all rights and interests arising prior to the effective date of any such denunciation.

Article XLVII—Review Conferences, amendments and related matters

1. – The Depositary, in consultation with the Supervisory Authority, shall prepare reports yearly, or at such other time as the circumstances may require, for the States Parties as to the manner in which the international regimen established in the Convention as amended by this Protocol has operated in practice. In preparing such reports, the Depositary shall take into account the reports of the Supervisory Authority concerning the functioning of the international registration system.
2. – At the request of not less than twenty-five per cent of the States Parties, Review Conferences of the States Parties shall be convened from time to time by the Depositary, in consultation with the Supervisory Authority, to consider:
   (a) the practical operation of the Convention as amended by this Protocol and its effectiveness in facilitating the asset-based financing and leasing of the assets covered by its terms;
   (b) the judicial interpretation given to, and the application made of the terms of this Protocol and the regulations;
   (c) the functioning of the international registration system, the performance of the Registrar and its oversight by the Supervisory Authority, taking into account the reports of the Supervisory Authority; and
   (d) whether any modifications to this Protocol or the arrangements relating to the International Registry are desirable.
3. – Any amendment to this Protocol shall be approved by at least a two-thirds majority of States Parties participating in the Conference referred to in the preceding paragraph and shall then enter into force in respect of States Parties which have ratified, accepted or approved such amendment when it has been ratified, accepted or approved by ten States Parties in accordance with the provisions of Article XXXVIII relating to its entry into force.

Article XLVIII—Depositary and its functions

1. – Instruments of ratification, acceptance, approval or accession shall be deposited with the International Institute for the Unification of Private Law (UNIDROIT), which is hereby designated the Depositary.
2. – The Depositary shall:

(a) inform all Contracting States of:
   (i) each new signature or deposit of an instrument of ratification, acceptance, approval or accession, together with the date thereof;
   (ii) the date of entry into force of this Protocol;
   (iii) each declaration made in accordance with this Protocol, together with the date thereof;
   (iv) the withdrawal or amendment of any declaration, together with the date thereof; and
   (v) the notification of any denunciation of this Protocol together with the date thereof and the date on which it takes effect;
(b) transmit certified true copies of this Protocol to all Contracting States;
(c) provide the Supervisory Authority and the Registrar with a copy of each instrument of ratification, acceptance, approval or accession, together with the date of deposit thereof, of each declaration or withdrawal or amendment of a declaration and of each notification of denunciation, together with the date of notification thereof, so that the information contained therein is easily and fully available; and
(d) perform such other functions customary for depositaries.

IN WITNESS WHEREOF the undersigned Plenipotentiaries, having been duly authorised, have signed this Protocol.

DONE at Berlin, this ninth day of March, two thousand and twelve, in a single original in the English and French languages, both texts being equally authentic, such authenticity to take effect upon verification by the Secretariat of the Conference under the authority of the President of the Conference within ninety days hereof as to the consistency of the texts with one another.

Bibliography

Aganaba, T., Assessing the Support for the Space Assets Protocol to the UNIDROIT Cape Town Convention, Report of International Institute of Space Commerce (2009), available at http://www.iisc.im/documents/IISC_-_09_-_P.12.1%5B2%5D_1.pdf.

Akseli, O., *International Secured Transactions Law: Facilitation of Credit and International Conventions and Instruments* (2011).

Aoki, S., "In Search of the Current Legal Status of the Registration of Space Objects", in *2010 Proceedings of the International Institute of Space Law* 245 (2011).

Balsano, A.M., "Modernising Secured Transactions Law: A Viewpoint from the European Space Agency", 8 Unif. L. Rev./Revue de droit uniforme 375 (2003).

Barrett, R.J., "Outer Space and Air Space: The Difficulties in Definition", Air University Review (1973), available at www.airpower.au.af.mil/airchronicles/aureview/1973/May-Jun/barrett .html.

Baumann, I. and Dodel, H., "The ITU Filing of Satellite Systems", in Smith, L.J., and Baumann, I. (eds.), *Contracting for Space* at 367 (2011).

Bertrand, T., "Le financement des satellites et les sûretés réelles mobilières conventionnelles." Mémoire pour l'obtention du Diplôme d'Etude Approfondie (DEA) de Droit International Public et Privé, Nice, Université de Nice Sophia-Antipolis, Institut du droit de la paix et du développement (IDPD) (2000).

Blount, P.J., "Jurisdiction in Outer Space: Challenges of Private Individuals in Space", 33 J. Space L. 299 (2007).

Bohlmann, U., "Short Report on the Background and Current State of Play of the UNIDROIT Convention on International Interest in Mobile Equipment and Its Preliminary Draft Protocol on Matters Specific to Space Assets", 38 ECSL News 22 (2011).

Bollweg, H.-G., "Diplomatic Conference for the Adoption of the Draft Protocol to the Convention on International Interests in Mobile Equipment on Matters Specific to Space Assets in Berlin from February 27 to March 9, 2012", 60 Zeitschrift für Luft- und Weltraumrecht 692 (2011).

Bollweg H.-G., and Gerhard, M., "Sicherungsrechte an Luftfahrtausrüstung und Weltraumeigentum – Die Entwürfe einer UNIDROIT/ICAO – Konvention über Internationale Sicherungsrechte an beweglicher Ausrüstung und der Protokolle über Luftfahrtausrüstung und Weltraumeigentum", 50 Zeitschrift für Luft- und Weltraumrecht 373 (2001).

Bourbonniere, M., "National-Security Law in Outer Space: The Interface of Exploration and Security", 70 J. Air L. & Com. 3 (2005).

Bourély, M. "Rules of International Law Governing the Commercialisation of Space Activities", in *Proceedings of the 29th Colloquium on the Law of Outer Space* 157 (1987).

Bridge, R.L., "International Law and Military Activities in Space," 13 Akron L. Rev. 649 (1979).

Buxbaum, H.L., "Unification of the Law Governing Secured Transactions: Progress and Prospects for Reform", 8 Unif. L. Rev./Revue de droit uniforme 321 (2003).

Campbell, D. (ed.), *International Secured Transactions* (2003).

Chatterjee, J., "Reflections on the Concepts of 'Liability' and 'Jurisdiction' in a Privatised Atmosphere, in Sundahl, M. and Gopalaskrishnan, V. (eds.), *New Perspectives on Space Law* at 183 (2011).

Chatzipanagiotis, M., "Registration of Space Objects and Transfer of Ownership in Orbit", 56 Zeitschrift für Luft- und Weltraumrecht 229 (2007).
——, *The Legal Status of Space Tourists in the Framework of Commercial Suborbital Flights* (2011).
Cheng, B., "The Extra-Terrestrial Application of International Law", 18 Current Legal Problems 132 (1965).
——, "The Legal Status of Outer Space and Relevant Issues: Delimitation of Outer Space and Definition of Peaceful Use", 11 J. Space L. 89 (1983).
Chinkin, C., and Kessedjian, C., "The Legal Relationship Between the Proposed UNIDROIT Convention and Its Equipment-specific Protocols", 4 Unif. L. Rev/Revue de droit uniforme 323 (1999).
Christol, C.Q., *Space Law: Past, Present, and Future* (1991).
Clark, L.S., "The 2001 Cape Town Convention on International Interests in Mobile Equipment and Aircraft Equipment Protocol: Internationalising Asset-Based Financing Principles for the Acquisition of Aircraft and Engines", 69 J. Air L. & Com. 3 (2004).
Cuming, R.C.C, "Study of International Regulation of Aspects of Security Interests in Mobile Equipment", 1 Unif. L. Rev/Revue de droit uniforme 63 (1990).
——, "The Draft UNIDROIT Convention on International Interests in Mobile Equipment", 30 UCC L.J. 365 (1998).
——, "'Hot Issues' in the Development of the (Draft) Convention International Interests in Mobile Equipment and the (Draft) Aircraft Equipment Protocol", 34 Int'l Law. 1093 (2000).
Cunningham, R.D., "Space Commerce and Secured Financing – New Frontiers for the U.C.C.", 40 Bus. Law. 803 (1985).
Dahan, F. and Simpson, J. (eds.), *Secured Transactions Reform and Access to Credit* (2008).
de la Peña, N., "Reforming the Legal Framework for Security Interests in Mobile Property", 4 Unif. L. Rev./Revue de droit uniforme 347 (1999).
Davis, S.A., "Unifying the Final Frontier: Space Industry Financing Reform", 106 Commercial Law Journal 455 (2001).
Dempsey, P. (ed.), *Space Law* (2012).
Diedericks-Verschoor, I.H. Ph., "The Development of Financing of Spacecraft", in *Proceedings of the Fortieth Colloquium on the Law of Outer Space* 212 (1998).
——, "Financing and Insurance Aspects of Spacecraft", 24 J. Space L. 97 (1996).
Diederiks-Verschoor, I.H.Ph., and Kopal V., *An Introduction to Space Law* (3rd ed. 2008).
Dodge, M.S., "Sovereignty and the Delimitation of Airspace: A Philosophical and Historical Survey", 35 J. Space L. 5 (2009).
dos Santos, A.F., "Developing Countries and the UNIDROIT Protocol on Space Property", in *Proceedings of the Forty-Fifth Colloquium on the Law of Outer Space* 23 (2003).
——, "Financing of Space Assets", 19 Space Policy 127 (2003).
Farimiñán Gilbert, J.M. de, "Private International Law and Space Activities", 21 ECSL News 4 (2000).
Farimiñán Gilbert, J.M. de, and Rodriguez Serrano, V., "Derecho Internacional Privado y Actividades Espaciales", Boletín n° 11 del Centro Española de derecho especial (2001).
Farimiñán Gilbert, J.M. de, and Caparrós del Moral, R., "Análisis general del Convenio de UNIDROIT relativo a las garantías internacionales sobre bienes de equipo móvil: Especial referencia al Proyecto preliminar de Protocolo conexo sobre cuestiones concretas relacionadas con la propiedad espacial", 24 Revista Española de Derecho Internacional 501 (2003).
Fernandez-Brital, O., "UN COPUOS Legal Subcommittee March-April 1993 Meeting and the Limit Between Air Space and Outer Space", in *Proceedings of the Thirty-Seventh Colloquium on the Law of Outer Space* 244 (1995).
Feyte, C., "Le projet de Convention Unidroit relative aux garanties internationales portant sur des objets mobiles: L'irrésistible ascension du droit international privé dans la régulation des activités spatiales", 21 ECSL News 3 (2000).

Fontmichel, A. de, "Commentaires sur 'l'avant-projet de Protocole sur les questions spécifiques aux matériels d'équipement spatial au projet de Convention d'Unidroit relative aux garanties internationales portant sur des matériels d'équipement mobiles'", 50 Zeitschrift für *Luft-* und *Weltraumrecht* 527 (2001).
Forkosch, M.D., *Outer Space and Legal Liability* (1982).
Forster, I., "The UNIDROIT Convention on International Interests in Mobile Equipment and the Preliminary Draft Protocol on Matters Specific to Space Assets: What Should Be the Position of Germany in Respect of the Protocol and in Particular the Various Opt-in/opt-out Alternatives Considering the Commercial Needs and Legal Implications?," in *Proceedings of the Forty-Sixth Colloquium on the Law of Outer Space* 48 (2004).
Gál, G., *Space Law* (1969).
Galloway, E., "The Definition of Outer Space", in *Proceedings of the Tenth Colloquium on the Law of Outer Space* 268 (1968).
Garcia, J.L., "Dispute Resolution in Matters Related to International Interests in Space Assets", 12 Unif. L. Rev./Revue de droit uniforme 323 (2007).
Gerhard, M., "Transfer of Operation and Control with Respect to Space Objects – Problems with Responsibility and Liability of States", 51 Zeitschrift für Luft- und Weltraumrecht 571 (2002).
Gerhard, M. and Gungaphul-Brocard, K., "The Impact of National Space Legislation on Space Industry Contracts", in Smith, L.J., and Baumann, I. (eds.), *Contracting for Space* at 59 (2011).
Goedhart, R.F.A., *The Never Ending Dispute: Delimitation of Air Space and Outer Space* (1996).
Goldsmith, J.L. and Posner, E.A., "A Theory of Customary International Law", 66 Univ. Chicago Law Review 1113 (1999).
Goode, R., "Transcending the Boundaries of Earth and Space: The Preliminary Draft UNIDROIT Convention on International Interests in Mobile Equipment", 3 Unif. L. Rev./Revue de droit uniforme 52 (1998).
——, "Cape Town Convention on International Interests in Mobile Equipment: A Driving Force for International Asset-Based Financing", 7 Unif. L. Rev./Revue de droit uniforme 3 (2002).
——, *Convention on International Interests in Mobile Equipment and Protocol Thereto on Matters Specific to Aircraft Equipment: Official Commentary* (2nd ed. 2008).
Gopalan, S., "Harmonization of Commercial Law: Lessons from the Cape Town Convention on International Interests in Mobile Equipment", 9 Law & Bus. Rev. Am. 255 (2003).
Gorove, K.M., "Delimitation of Outer Space and the Aerospace Object – Where is the Law", 28 J. Space L. 11 (2000).
Gorove, S., "Liability in Space Law: An Overview", 8 Annals Air & Space L. 373 (1983).
——, "Toward a Clarification of the Term 'Space Object': An International Legal and Policy Imperative?", 21 J. Space L. 11 (1993).
——, "Aerospace – Legal and Policy Issues for Air and Space Law", 25 J. Space L. 101 (1997).
Hazan, N., "The UNIDROIT Preliminary Draft Protocol on Matters Specific to Space Assets", 28 Annals Air & Space L. 219 (2003).
Hermida, J., *Legal Basis for a National Space Legislation* (2004).
——, "Transfer of Satellites in Orbit: An International Law Approach", in *Proceedings of the Forty-Sixth Colloquium on the Law of Outer Space* 189 (2004).
Hertzfeld, H.R. and von der Dunk, F.G., "Bringing Space Law into the Commercial World: Property Rights Without Sovereignty", 6 Chi. J. Int'l L. 81 (2005).
Hobe, S., Schmidt-Tedd, B. and Schrogl, K. (eds.), *Cologne Commentary on Space Law, Vol. 1* (2009).
Höpfli, M., "Rechtsrahmen für die wirtschaftliche Nutzung des Weltraumes durch Privatunternehmen: Eine Analyse zentraler Rechtsfragen der wirtschaftlichen Nutzung des Weltraumes durch Privatunternehmen unter besonderer Berücksichtigung der *Convention on International Interests in Mobile Equipment* vom November 2001 in Verbindung mit dem UNIDROIT *Preliminary Draft Protocol on Matters Specific to Space*

Assets vom Januar 2002, Würzburg, 2003", Universität Würzburg, Diss. (Würzburger rechtswissenschaftliche Schriften, 42) (2002).

Hörl, K.-U., and Gungaphul, K., "Problems Related to 'Change of Ownership' with Respect to Registration – The Industry View", in *Proceedings of the Project 2001 Plus Workshop "Current Issues in the Registration of Space Objects"* 63 (2005).

Hörl, K.-U., and Hermida, J., "Change of Ownership, Change of Registry? Which Objects to Register, What Data to be Furnished, When and Until When?", in *Proceedings of the IISL/ ECSL Symposium* 454 (2003).

Hudson, J., "The Case Against Secured Lending", 15 International Review of Law and Economics 47 (1995).

Hurwitz, B.A., *State Liability for Outer Space Activities* (1992).

Jasentuliyana, N., and Lee, R.S.K. (eds.), *Manual on Space Law, Vol. I* (1979).

Jakhu, R.S. (ed.), *National Regulation of Space Activities* (2010).

Kaul, R., "Control of Space Assets: Ethics in International Diplomacy", 31 Annals Air & Space L. 447 (2006).

Kerrest, A., "Launching Spacecraft from the Sea and the Outer Space Treaty: The Sea Launch Project," in *Proceedings of the 40th Colloquium on the Law of Outer Space* 264 (1997).

——, "Remarks on the Responsibility and Liability for Damage Caused by Private Activity in Outer Space", in *Proceedings of the 40th Colloquium on the Law of Outer Space* 134 (1998).

——, "Remarks on the Notion of the Launching State", in *Proceedings of the Forty-Second Colloquium on the Law of Outer Space* at 308 (1999).

Klucka, J., "The Role of Private International Law in the Regulation of Outer Space", 39 Int'l & Comp. Law Quarterly 918 (1990).

Kopal, V., "What is 'Outer Space' in Astronautics and Space Law", in *Proceedings of the Tenth Colloquium on the Law of Outer Space* 275 (1968).

——, "The Question of Defining Outer Space", 8 J. Space L. 154 (1980).

Kozuka, S., "The Commercialisation of the Space and the Legal Infrastructure for Asset Based Financing: Commentary on the Draft Protocol on Space Assets to the Cape Town Convention (Parts 1)", 46 Sophia L. Rev. 1 (2003).

——, "The Commercialisation of the Space and the Legal Infrastructure for Asset Based Financing: Commentary on the Draft Protocol on Space Assets to the Cape Town Convention (Part 2)", 47 Sophia L. Rev. 1 (2003).

Kozuka, S. and Taniguchi, F., "The Economic Assessment of the Space Assets Protocol to the Cape Town Convention", in *2011 Proceedings of the International Institute of Space Law* 409 (2012).

Kreisel, J., and Lee, B.H., "Space Entrepreneurship – Status & Prospects", in Schrogl, K.-U., Mathieu, C., and Peter, N. (eds.), *Yearbook on Space Policy 2006/2007: New Impetus for Europe*, at 254 (2008).

Kronke, H., "The Draft UNIDROIT Convention on International Interests in Mobile Equipment and the Draft Protocol on Matters Specific to Space Assets", in Böckstiegel, K.H. (ed.), *"Project 2001" – Legal Framework for the Commercial Use of Outer Space: Recommendations and Conclusions to Develop the Present State of the Law*, at 643 (2002).

Larsen, P.B., "Creditors' Secured Interests in Satellites", in *Proceedings of the Thirty-Fourth Colloquium on the Law of Outer Space* 233 (1991).

Larsen, P.B., and Heilbock, J.A., "UNIDROIT Project on Security Interests: How the Project Affects Space Objects", 64 J. Air L. & Com. 703 (1999).

Larsen, P.B., "Future Protocol on Security Interests in Space Assets", 67 J. Air L. & Com. 1071 (2002).

——, "UNIDROIT Space Protocol: Comments on the Relationship between the Protocol and Existing International Space Law, in *Proceedings of the Forty-Fourth Colloquium on the Law of Outer Space* 187 (2002).

——, "Critical Issues in the UNIDROIT Draft Space Protocol", in *Proceedings of the Forty-Fifth Colloquium on the Law of Outer Space* 2 (2003).

——, "Space Asset Financing and Trade Issues", in *Proceedings of the Forty-Sixth Colloquium on the Law of Outer Space* 228 (2004).

——, "Brief Report on the Final Space Protocol", available at www.iislweb.org/docs/2012_unidroit.pdf.

——, "The Draft Space Protocol and Jurisdiction over Commercial Space Assets", in *2011 Proceedings of the International Institute of Space Law* 485 (2012).

Lawrence, L., *Lawrence's Anderson on the Uniform Commercial Code* (2012).

Lee, R.J., "Reconciling International Space Law with the Commercial Realities of the Twenty-First Century", 4 Singapore Journal of Int'l & Comp. Law, 194, 200–201 (2000).

——, "Effects of Satellite Ownership Transfers on the Liability of the Launching States", in *Proceedings of the Forty-Third Colloquium on the Law of Outer Space* 148 (2001).

Lochan, R., "Cape Town Convention & Space Protocol: A Critical Analysis", in Gopalakrishnan, V. and Lochan, R. (eds.), *Conference Proceedings of the ISRO/IISL Space Law Conference 2005: Bringing Space Benefits to the Asian Region*, 6–39 (2006).

Lyall, F., and Larsen, P.B., *Space Law: A Treatise* (2009).

Magnussen, A., *The Communications Satellite as Collateral: Swedish Property Law in Outer Space (Juridiska Fakulteten Skriftserien Nr. 58)*, at 92 (1999).

Marchisio, S., "Le protocole spatial d'Unidroit", 12 Géo Observateur 3 (2002).

——, "The Evolutionary Stages of the Legal Subcommittee of the United Nations Committee on the Peaceful Uses of Outer Space (COPUOS)", 31 J. Space L. 219 (2005).

——, "The UNIDROIT Draft Protocol on Space Assets: Towards the Diplomatic Conference?", 37 ECSL News (2010).

McCormack, G., *Secured Credit Under English and American Law* (2004).

——, *Secured Credit and the Harmonization of Law: The UNCITRAL Experience* (2011).

McGairl, S.J., "The Proposed UNIDROIT Convention: International Law for Asset Finance (Aircraft)", 4 Unif. L. Rev./Revue de droit uniforme 439 (1999).

Meisner, S.M., "Global Telecommunications Competition a Reality: United States Complies with WTO Pact", 13 American Univ. Int'l L. Rev. 1345 (1998).

Mello, M., "Hagan v. Australia: A Sign of the Emerging Notion of Hate Speech in Customary International Law", 28 Loyola of Los Angeles Int'l & Comp. Law Review 365, 368 (2006).

Meredith, Pamela, "The Space Protocol: Provisions for Insurer Salvage", UNIDROIT Doc. DCME-SP – Doc. 4 (July 2011).

Miller, F.H. (ed.) *Hawkland's Uniform Commercial Code Series* (2012).

Morgan, R.A., "Military Use of Commercial Communication Satellites: A New Look at the Outer Space Treaty and 'Peaceful Purposes'", 60 J. Air L. & Com. 237, 242 (1994).

Mosteshar, Sa'id, "Financing Space Assets: The Unidroit Solution Examined," available at www.esoa.net/upload/files/news/unidroit/analysisoflegalissues.pdf.

Nesgos, P., "New Developments in Space Law Concerning Financing and Risk Management", 27 Annals Air & Space L. 477 (2002).

Nesgos, P., "Satellites and Transponders", in Wong, J. and Dubin, B. (eds.) *Equipment Leasing* (2005).

Ospina, S., "El Protocolo sobre bienes espaciales de UNIDROIT: ¿Un paso hacia la privatización del Espacio Ultraterrestre?", 22 CIDA-E: revista del Centro de Investigación y Difusión Aeronáutica-Espacial 27 (2002).

——, "The Concepts of Assets and Property: Similarities and Differences, and Their Applicability to Undertakings in Outer Space", in *Proceedings of the Forty-Fifth Colloquium on the Law of Outer Space* 12 (2003).

Oyekunle, T., "Draft Protocol on Matters Specific to Space Assets to the Convention on International Interests in Mobile Equipment", in *Proceedings of the United Nations/Nigeria Workshop on Space Law: Meeting International Responsibility and Addressing Domestic Needs, Abuja, 21–24 November 2005* 519 (2006).

Panahy, D.A., and Mittal, R., "The Prospective UNIDROIT Convention on International Interests in Mobile Equipment as Applied to Space Property", 4 Unif. L. Rev./Revue de droit uniforme 303 (1999).

Peeters, W., Gurtuna, O., and Zervos, V., "Challenges and Opportunities in Space Financing for Emerging Entrepreneurs", in *Proceedings of the 59th International Astronautical Congress, Vol. VI* 9618 (2008).

Perek, L., "Scientific Criteria for the Delimitation of Outer Space", 5 J. Space L. 111 (1977).

Porras, D.A., "The Need for Timeous Completion of the Preliminary Draft Protocol to the Cape Town Convention on Matters Specific to Space Assets", 22 ESPI Persp. (2009).

Poulain, B., "L'avant-projet de Protocole portant sur les questions spécifiques aux biens spatiaux à la Convention relative aux garanties internationales portant sur des matériels d'équipement mobiles", UNIDROIT, avaiable at www.unidroit.org/english/conventions/mobile-equipment/bibliography/articles/space/poulain.pdf.

Qizhi, H., "The Problem of Definition and Delimitation in Outer Space", 10 J. Space L. 157 (1982).

Raustiala, K., "The Geography of Justice", 73 Fordham Law Review 2501 (2005).

Reif, S.U., "Shaping a Legal Framework for the Commercial Use of Outer Space.

Recommendations and Conclusions from Project 2001", 18 Space Pol'y 157 (2002).

Reisman, W.M. (ed.), *Jurisdiction in International Law* (1999).

Ribbelink, O.M., "The Draft Protocol on Matters Specific to Space Assets and the (Unidroit) Convention on International Interests in Mobile Equipment", in *Proceedings of the United Nations/International Institute of Air and Space Law Workshop: Capacity Building in Space Law, The Hague, 18–21 November 2002* 84 (2003).

——, "The Protocol on Matters Specific to Space Assets", 12 European Rev. Private L. 37 (2004).

Rosenfield, S.B., "Where Air Space Ends and Outer Space Begins", 7 J. Space L. 137 (1979).

Ryngaert, C., *Jurisdiction in International Law* (2008).

Sanchez-Aranzamendi, M., "Economic Impacts of National Space Legislation and the Establishment of Fair Conditions for Commercial Activities", PowerPoint presentation given at the IISL-ECL Symposium 2010, UNCOPUOS Legal Subcommittee (Mar. 22, 2010).

Saunders, A., Srinivasan, A., Walter, I., and Wool, J., "The Economic Implications of International Secured Transactions Law Reform: A Case Study", 20 Univ. of Pennsylvania. J. Int'l Econ. L. 309 (1999).

Saunders, A., and Walter, I., "Proposed UNIDROIT Convention on International Interests in Mobile Equipment as Applicable to Aircraft Equipment Through the Aircraft Equipment Protocol: Economic Impact Assessment", 23 Air & Space L. 339 (1998).

Schmidt-Tedd, B., and Arnold, I., "The Unidroit Draft of a Space Assets Protocol: A Civil Law Instrument under a Public Framework", in *Proceedings of the Fifty-First Colloquium on the Law of Outer Space* 71 (2009).

Schrogl, K. and Davies, C., "A New Look at the 'Launching State": The Results of the UNCOPUOS Legal Subcommittee Working Group 'Review of the Concept of the Launching State 2000–2002'," in *Proceedings of the 45th Colloquium on the Law of Outer Space* 286 (2002).

Smith, L.J., and Baumann, I. (eds.), *Contracting for Space* (2011).

Soriano, M.V. et al., *The European Insolvency Regulation: Law and Practice* (2004).

Stanford, M.J., "Completion of a First Draft of UNIDROIT's Planned Future Convention on International Interests in Mobile Equipment", 1 Unif. L. Rev./Revue de droit uniforme 274 (1996).

——, "A Broader or Narrower Band of Beneficiaries for the Proposed New International Regimen?: Some Reflections on the Merits of the Convention/Protocol Structure in Facilitating the Former", 4 Unif. L. Rev./Revue de droit uniforme 242 (1999).

——, "The UNIDROIT Project for the Creation of a New Legal Regimen Governing the Taking of Security in High-Value Mobile Assets: A Window of Opportunity for the Commercialisation of Space", in *Proceedings of the Third ECSL Colloquium: International Organisations and Space Law, Perugia, Italy, 6–7 May, 1999* 375 (1999).

——, "Commentary Paper: UNIDROIT's Project for the Creation of a New Regimen Governing the Taking of Security in High-Value Mobile Assets: A Window of Opportunity in the Context of the Privatisation and Commercialisation of Space", in *Proceedings of the Workshop on Space Law in the Twenty-first Century, organised by the International Institute of Space Law with the United Nations Office for Outer Space Affairs, UNISPACE III Technical Forum, July 1999* 148 (1999).

——, "UNIDROIT's Project for the Creation of a New International Regimen Governing the Taking of Security over High-Value Mobile Equipment: A Window of Opportunity for the Financing of Railway Rolling Stock and Commercial Space Activities", 13 Revista mexicana de derecho internacional privado y comparado 91 (2003).

——, "The Preliminary Draft Protocol to the Cape Town Convention on Matters Specific to Space Assets: A Unique Opportunity to Expand the Benefits of Space-Based Services and to Broaden the Market for Commercial Space Activities in General", in *Proceedings of the United Nations/Islamic Republic of Iran Workshop on Space Law: Role of International Space Law in the Development and Strengthening of International and Regional Cooperation in the Peaceful Exploration and Use of Outer Space, Tehran, 8–11 November 2009* (2010). Stanford, M.J., "The Availability of a New Form of Financing for Commercial Space Activities: The Extension of the Cape Town Convention to Space Assets," 1 Cape Town Convention J. 109 (2012).

Stanford, M.J., and Fontmichel, A. de, "Overview of the Current Situation Regarding the Preliminary Draft Space Property Protocol and Its Examination by COPUOS", 6 Unif. L. Rev./ Revue de droit uniforme 60 (2001).

Stanford, M.J., and Poulain, B., "The Preliminary Draft Space Assets Protocol to the Cape Town Convention on International Interests in Mobile Equipment", 26 ECSL News 4 (2003).

——, "The Preliminary Draft Space Assets Protocol to the Convention on International Interests in Mobile Equipment on Matters Specific to Space Assets", in *Proceedings of the United Nations/Thailand/European Space Agency Workshop on Space Law: Activities of States in Outer Space in Light of New Developments: Meeting International Responsibilities and Establishing National Legal and Policy Frameworks, Bangkok, 16–19 November 2010* (2010).

Sterns, P.M., and Tennen, L.I., "Doing Business in Space: Operating Strategies for a Changing Market", in *Proceedings of the Twenty-Ninth Colloquium on the Law of Outer Space* 183 (1987).

——, "Security Interests and Creditor's Remedies in the Law of Outer Space", in *Proceedings of the Thirty-Third Colloquium on the Law of Outer Space* 102 (1990).

Stewart, Jr., J.T., "Should There Be a Mortgage Convention for Space Activity Investors?", in *Proceedings of the Twenty-Fifth Colloquium on the Law of Outer Space* 251 (1983).

Sundahl, M.J., "Unidentified Orbital Debris: The Case for a Market-Share Liability Regime", 24 Hastings Int'l & Comp. L. Rev. 125 (2000).

——, "The 'Cape Town Approach': A New Method of Making International Law", 44 Colum. J. Transnat'l L. 339 (2006).

——, "The Cape Town Convention: Unleashing Capital for the Space Industry", in Gopalakrishnan, V. and Lochan, R. (eds.), *Conference Proceedings of the ISRO/IISL Space Law Conference 2005: Bringing Space Benefits to the Asian Region* 6–31 (2006).

——, "Information Warfare: The Legal Aspects of Using Satellites and Jamming Technologies in Propaganda Battles", in *Proceedings of the 49th Colloquium on the Law of Outer Space* (2007).

——, "Iraq, Secured Transactions and the Promise of Islamic Law", 40 Vanderbilt Journal of Transnational Law 1301 (2007).

——, "The Duty to Rescue Space Tourists and Return Private Spacecraft", 35 J. Space L. 163 (2009).

——, "Space Tourism and Export Controls: A Prayer for Relief, 75 Journal of Air Law & Commerce 581 (2010).

Tronchetti, F., "The Future of the UNIDROIT Draft Space Protocol: Legal Proposals to Favour and Stimulate Its Success", in *2008 Proceedings of the International Institute of Space Law* 60 (2009).
van Fenema, P., "The UNIDROIT Space Protocol, the Concept of 'Launching State', Space Traffic Management and the Delimination of Outer Space (Report of the 41st Session of the UNCOPUOS Legal Subcommittee)", 27 Air & Space L. 266 (2002).
Vereshchetin, V.S., "The Law of Outer Space in the General Legal Field (Commonality and Particularities", in *Proceedings of the Fifty-Second Colloquium on the Law of Outer Space* 3 (2010).
——, "Space Activities of 'Nongovernmental Entities': Issues of International and Domestic Law", in *Proceedings of the Twenty-Sixth Colloquium on the Law of Outer Space* 261 (1983).
von der Dunk, F.G., "Liability versus Responsibility in Space Law: Misconception or Misconstruction?", in *Proceedings of the Thirty-Fourth Colloquium on the Law of Outer Space* 363 (1991).
——, "The Illogical Link: Launching, Liability and Leasing", in *Proceedings of the Thirty-Sixth Colloquium on the Law of Outer Space* 349 (1993).
——, "Public Space Law and Private Enterprise – The Fitness of International Space Law Instruments for Private Space Activities", in *Legal Framework for Privatising Space Activities, Proceedings of the Project 2001 Workshop on Legal Issues of Privatising Space Activities*, at 12 (1999).
von Mehren, A.T., *Adjudicatory Jurisdiction: General Theories Compared and Evaluated*, 63 Boston Univ. Law Review 279 (1983).
Wirin, W.B., "Practical Implications of Launching State – Appropriate State Definitions", in *Proceedings of the 37th Colloquium on the Law of Outer Space* 109 (1994).
Wood, P.R., *Comparative Law of Security Interests and Title Finance* (2007).
Wool, J., "Rethinking the Notion of Uniformity in the Drafting of International Commercial Law: A Preliminary Proposal for the Development of a Policy-based Unification Model", 2 Unif. L. Rev./Revue de droit uniforme 46 (1997).
——, "The Case for a Commercial Orientation to the Proposed Unidroit Convention as Applied to Aircraft Equipment", 31 Law & Pol'y Int'l Bus. 79 (1999).
——, "The Next Generation of International Aviation Finance Law: An Overview of the Proposed UNIDROIT Convention on International Interests in Mobile Equipment as Applied to Aircraft Equipment", 20 U. Pa. J. Int'l Econ. L. 499 (1999).
Xinmin, M.A., "Statement on International Co-operation in the Peaceful Uses of Outer Space", 6 Chinese J. Int'l L. 507 (2007).
Yun Z., "Revisiting Selected Issues in the Draft Protocol to the Cape Town Convention on Matters Specific to Space Assets", 76 J. Air L. & Com. 805 (2011).
Zhukov, G.P., "The Problem of the Definition of Outer Space", in *Proceedings of the Tenth Colloquium on the Law of Outer Space* 271 (1968).
Zhukov, G.P. and Kolosov, Y., *International Space Law* (1984).

Index

www.ingramcontent.com/pod-product-compliance
Lightning Source LLC
LaVergne TN
LVHW020652300926
849161LV00004B/907
* 9 7 8 9 0 0 4 2 0 8 9 1 9 *